IN THIS TOGETHER

IN THIS TOGETHER

Connecting with Your Community to Combat the Climate Crisis

Marianne E. Krasny

COMSTOCK PUBLISHING ASSOCIATES
AN IMPRINT OF
CORNELL UNIVERSITY PRESS ITHACA AND LONDON

First published 2023 by Cornell University Press
Printed in the United States of America

Library of Congress Cataloging-in-Publication Data

Names: Krasny, Marianne E., author.
Title: In this together : connecting with your community to combat the climate crisis / Marianne E. Krasny.
Description: Ithaca [New York] : Comstock Publishing Associates, an imprint of Cornell University Press, 2023. | Includes bibliographical references and index.
Identifiers: LCCN 2022022656 (print) | LCCN 2022022657 (ebook) | ISBN 9781501768590 (paperback) | ISBN 9781501768583 (pdf) | ISBN 9781501768576 (epub)
Subjects: LCSH: Climate change mitigation—Citizen participation. | Environmental protection—Citizen participation. | Environmental policy—Citizen participation. | Climatic changes—Effect of human beings on.
Classification: LCC TD171.75 .K73 2023 (print) | LCC TD171.75 (ebook) | DDC 363.738/74—dc23/eng/20221011
LC record available at https://lccn.loc.gov/2022022656
LC ebook record available at https://lccn.loc.gov/2022022657

Dedicated to the stalwart volunteers at Elders Climate Action who brought me into the world of climate activism, including Mark Cook, Gerri Friedman, Frances Stewart, and Leslie Wharton, as well as our fearless director of operations, Jen Chandler

Contents

Preface

Maybe it's too late. Scientists say we have just a few years to avert the worse. Yet I, like so many others, nevertheless feel compelled to do *something* to push back against climate catastrophe. We want to take actions that make a real difference, that contribute meaningfully to a livable world for ourselves, our children, our grandchildren, our friends. But what could those actions possibly be?

These were the thoughts swirling around my mind after the Intergovernmental Panel on Climate Change (IPCC) report, *Global Warming of 1.5°C*, hit the media in fall 2018. My time to make the world a better place was running out. And the tiny actions that I could take on my own—eat less meat, reduce food waste, turn down the heat, avoid air travel—all felt just that: tiny. What real difference could I make in the face of the megafires devouring the California landscape? What could I do to stop the blooms of harmful algae in nearby Cayuga Lake or along the coasts of Florida, as waves of climate disruption engulfed our lives?

I pondered these thoughts during October, then November, then December of 2018. I read about individual behaviors versus collective action. Was there a way I could foment collective action—something bigger than what I could do alone?

This is when I hit on the idea of network climate action. Network climate action has three steps: (1) identify the most effective actions to draw down greenhouse gas emissions, (2) take those actions yourself, and (3) think of creative and fun ways to draw your family and friends—that is, your tight social network—into taking those actions alongside you. We can involve family and friends in a range of actions, from things we do at home, such as eating less meat or composting food waste, to actions aimed at changing policies, such as donating to advocacy organizations and sending messages to legislators.

This book is written for the average person who wants to do something meaningful about climate change. Rather than tell stories of climate heroes, I have trained my sights on regular people such as myself and my friends. Rather than offer a list of fifty things you can do to improve the environment, I focus on a handful of the most impactful actions you can take. I show how we can scale up those actions through our family, friends, workplace, and volunteer organizations. And I share my journey—how I found meaningful answers to my questions about individual responsibility, action, and climate—in the hope that reading about my path will help you to chart your own. Each of us can find a role in living a climate-friendly life and influencing climate policy and, together with our friends and family who join us, can push back against the rising tide of climate catastrophe.

Acknowledgments

I would like to thank my editor, Emily Hopkins, who helped me through a significant revision of the original text and who is also a skilled cartoonist. Emily drew the cartoons scattered throughout the book that help convey the messages with a smack of humor. I am also deeply indebted to Kitty Liu at Cornell University Press for championing this project through multiple jumps and starts, and to her colleagues at the press, including Jacqulyn Teoh, Mary Kate Murphy, and Don McKeon, who helped this book become a reality. Several friends, family members, and colleagues made comments on the manuscript, including Paul Cody, Danny Rosenberg Daneri, Ted Krasny, the Reverend Michael Malcom, Melanie Pinkert, and Bjorn Whitmore. Stuart Capstick and two anonymous reviewers also provided helpful comments. At Cornell University, I thank the members of the Civic Ecology Lab, including Anne Armstrong, Elena Dominguez Contreras, Bethany Jorgensen, Alex Kudryavtsev, Yue Li, Leo Louis, Melanie Quinones Santiago, Xoco Shinbrot, and Mi Yan, who discussed ideas about network climate action and reviewed drafts of the manuscript. I also garnered ideas and inspiration from the students in my Cornell classes and participants in our global online courses and fellowship, some of whose stories are recounted in this book. Climate Reality local chapter leaders Thomas Hirasuna and Diane Stefani, along with Brett Walter and Matthew Vollrath of Climate Action Now, have made meaningful activism possible for me and many others. Finally, I would like to thank my brother Fred Krasny and sister-in-law Betty Baer for their ongoing support and ideas, my daughter, Aleysia Whitmore, who checked in with me frequently, and my sons, Bjorn Whitmore and Sylvan Whitmore, who urged me on during our late-night dumpster-diving expedition.

IN THIS TOGETHER

INTRODUCTION

Every day on my walk up the hill to Cornell University, I engage in "plalking," the Swedish activity of picking up litter while walking.[1] I gather up Dunkin' Donuts and Starbucks cups, scraps of aluminum foil gnawed by squirrels, squished beer cans, cigarette butts, and even the occasional used condom. As I gather what others have carelessly tossed to the ground, my feelings are mixed: disgust at touching someone else's trash, satisfaction in doing the right thing, and, most of all, deep uncertainty about the meaning of my meager cleanup efforts.

But as I lean over to grab another red Solo cup, I think to myself that each piece of plastic I remove means one less grimy item to be washed through the sewer grates that drain into Fall Creek. One less colorful plastic fragment to be snatched up by a hungry fish or bird. Of course, I realize how ridiculous my plalking habit might appear, given that every flood and storm surge carries massive amounts of trash and debris into our rivers, lakes, and oceans. But each cigarette butt I grab brings a narrow ray of hope—that maybe my action will somehow make a difference and my beloved Fall Creek will retain its natural beauty a while longer.

In the end, my tiny efforts help me avoid despair, as do the small actions I take to address the climate crisis. "Despair is paralysis. It robs us of agency. It blinds us to our own power and the power of the earth," writes scholar Robin Kimmerer. She goes on to lament, "Environmental despair is a poison."[2] Kimmerer's words ring true, regardless of whether despair is about a littered landscape or a changing climate.

This book is for individuals who, like me, are grappling with that despair. It is for people who are seeking hope through wrestling with what individuals can realistically do to address the colossal climate crisis. Individual lifestyle actions—things we can do in our homes, on our way to work, and throughout our daily lives—are important, but we need to be informed about the actions we choose. Some actions, such as eating more plant-rich foods and reducing the amount of food we waste, curtail our greenhouse gas emissions a lot more than other actions, such as recycling. We also need to consider actions, such as writing to our representatives, aimed at changing climate policy.

But acting alone will not bring about the policy changes needed to transform our greenhouse-gas-gushing energy, transportation, and food systems. How can we bring about structural or systemic change? This entails expanding on what we do daily, such as eating meat-free meals, and occasional individual advocacy actions, such as voting and protesting. It entails connecting with others taking similar actions. It means supporting—through donations and volunteering—organizations whose members take collective action to advocate for equitable climate policies and production systems.

Whether we choose a direct consumer action to reduce our emissions, such as eating less meat, or an indirect collective action to influence policy, such as joining a group advocating for electric vehicle charging stations, we can scale up the impact of our actions. We can do this by persuading our friends, family, and colleagues—that is, our close social networks—to join in a plant-rich brunch, a neighborhood composting project, or a volunteer "tweetstorm" organized by the nonprofit Climate Reality Project. We will have a greater impact if we invite our social networks to participate in individual and collective climate actions alongside us.

Emphasizing individual actions such as reducing the food we waste in our homes can deflect attention from the fossil fuel companies—the multi-billion-dollar culprits who bear massive responsibility for the climate crisis.[3] Yet, done right, individual actions *can* lead to systemic change. Individual action and systemic change feedback one to another—they are connected. People who change what they eat not only influence their family and friends but also join in collective actions to transform food systems—actions such as buying local foods, joining boycotts of unsustainable products, and volunteering to transport uneaten food from restaurants to food kitchens. As people join in these collective actions, they become part of political and ethical consumer movements attempting to influence government policies and private business practices.[4] And, like me, they may even participate in the climate movement by joining climate organizations whose members tweet their representatives, phone-bank to turn out environmental voters, and otherwise collectively advocate for climate-friendly policies.

The Fossil Fuel Industry Blames Consumers

Big business has a long history of trying to convince us that we, the consumers, are to blame for problems caused by its products. Take the tobacco companies. When I was growing up, a battle was brewing between the government and the cigarette companies. As evidence mounted that smoking was bad for your health, the cigarette industry fought back with misinformation or, more precisely, with downright lies. Part of their public disinformation campaign, and their defense when facing lawsuits, was to frame smoking as an issue of freedom of choice and personal responsibility and to blame those who chose to smoke and ended up with cancer or heart disease. A Philip Morris executive summed up the argument: "It all comes down to the individual's right to make up his own mind and to take responsibility for his own actions."[5] The cigarette companies gave hundreds of thousands of dollars to the science-defying Heartland Institute, which promoted their disinformation campaign.[6]

Years later, the same Heartland Institute, this time in cahoots with Exxon Mobil and the coal industry, found itself once again on the wrong side of history. The Heartland Institute and its fossil fuel industry collaborators disseminated blatant falsehoods to discredit climate science. One campaign claimed, "The most prominent advocates of global warming aren't scientists. They are murderers, tyrants, and madmen." In one of its more outrageous stunts, the institute installed a billboard along a Chicago expressway equating people who believe in climate science with the notorious Unabomber killer.[7]

Needless to say, the Heartland Institute is a malevolent actor in not just the tobacco debacle but also in aggravating the climate crisis. Its fellow bad actors, the fossil fuel behemoths, have used stealthy diversion tactics to draw attention away from the fact that they are responsible for about three-quarters of US greenhouse gas emissions in any one year.[8] They took their cues from the fake Crying Indian campaign in the early 1970s, in which Coca-Cola and the plastic cup manufacturer Dixie hired an Italian American actor to dress in traditional Native American garb and to shed tears over a litter-strewn landscape. The campaign's motto, "People start pollution. People can stop it," told the public that individuals, not the plastics and beverage industries, were responsible for litter.[9] If only consumers recycled, it would solve the litter problem, industry claimed, while knowing full well that recycling was not economically viable.

As far back as 1965, the fossil fuel companies' own scientists reported that greenhouse gas emissions were rising to dangerous levels.[10] Yet the companies continued to expand their investment in fossil fuels and to mislead the public by sowing doubt about the certainty of human-caused global warming. At the same time, they touted a different certainty: an ever-increasing global energy demand

and the notion that these energy needs must be met through fossil fuels. The spotlight was on the individuals who needed the fossil fuels to carry on their daily lives, let alone have brighter futures. Because they used the energy, individuals were to blame for any pollution that resulted. Never once did energy companies allow that they themselves might bear responsibility for fossil fuel emissions or that they could change their business model from fossil fuel extraction to renewable energy. In fact, the companies spewed emissions, lies, and myths about the unreliability of renewable energy.[11]

Once it became clear that the science about global warming was indisputable, the oil giants switched their rhetoric from doubting the science to touting the risks—that is, the economic risks of higher gas prices. The companies mounted greenwashing campaigns such as BP's "Beyond Petroleum," while continuing rampant oil and gas development. In the mid-2000s, BP was one of the first to launch the notion of a "personal carbon footprint" to steer people away from thinking about industry's own carbon "pipeline footprint."[12]

More recent anticlimate campaigns have used bots to sow disinformation. During a normal climate news cycle in 2017, bots accounted for about a

"Hey, your straw is not recyclable."

quarter of all tweets mentioning climate change, with bot tweets more prevalent than human-generated tweets in disseminating fringe views denying climate science and the need for action. Although as of 2021 researchers had not determined the source of the bots, they suspected fossil fuel companies and oil-producing countries, which stand to benefit from the public's confusion about climate change.[13]

To top it all off, fossil fuel companies have recently begun presenting themselves as saviors of the environment through their support of research and technology. They tout their industry's good deeds, while not so subtly shifting the focus to customers and drivers: "We're supporting research and technology efforts, curtailing our own greenhouse gas emissions and helping customers scale back their emissions of carbon dioxide"; "By enabling cars and trucks to travel farther on a gallon of fuel, drivers . . . emit less carbon dioxide per mile." Even volunteer tree planters benefit from the fossil fuel industry's largesse: "We're pleased to extend our support of . . . American Forests . . . whose 'Global Releaf 2000' program is mobilizing people around the world to plant and care for trees."[14]

The Fight over Individual Action

> **Limiting climate change requires interventions at multiple levels.**
>
> —Kristian Nielsen et al., environmental psychologists, "How Psychology Can Help Limit Climate Change"

Given the fossil fuel industry's track record of lies, deceptions, and outsized emissions, it is not surprising that the preeminent climate scientist and activist Michael Mann questions a focus on individuals changing their behaviors. Although he himself doesn't eat meat and gets his energy from renewable sources, he is alarmed by how the fossil fuel emitters have mounted multi-million-dollar campaigns to deflect our attention from their polluting industries. Mann's line of argument is that by picking up litter or by other individual acts such as turning down the heat and turning off the lights, we are buying into a false narrative. The fossil fuel companies have duped us into thinking that everyday consumers are responsible for the climate crisis and that if we each consumed less, then the crisis would go away. Companies could continue to exploit oil and gas resources and fend off government regulations.[15]

Mann has other reasons to question an emphasis on individual action. He is concerned that those who adopt climate-friendly behaviors will be seen as aggravatingly "good" or "virtuous." They end up casting shame on others who still barbecue steaks and drive gas-guzzling pickup trucks, leading these supposed "sinners" to cling all the more tightly to their perceived right to do whatever they please. Mann claims that "when the climate discourse devolves into a shouting match over diet and travel choices, and becomes about personal purity, behavior-shaming, and virtue-signaling, we get a divided community unable to speak with a united voice. We lose. Fossil fuel interests win."[16]

Mann is also troubled by "moral license" and the "boomerang effect." After focusing on relatively ineffective actions such as recycling, we feel righteous and moral. As a result, we feel that we now have the license to avoid more costly actions, such as taking public transport or advocating for renewable energy. Our relatively trivial environmental behaviors, such as picking up litter, boomerang as we purchase a ticket for a luxury Caribbean cruise.

Mann sums up his concerns about individual consumer action in his 2021 book *The New Climate War*: "Consumer choice doesn't build high-speed railways, fund research and development in renewable energy, or place a price on carbon emissions."[17] As the philosopher Walter Sinnott-Armstrong declared, "global warming is such a large problem that it is not individuals who cause it or who need to fix it. . . . Finding and implementing a real solution is the task of governments. Environmentalists should focus their efforts on those who are not doing their job rather than on those who take Sunday afternoon drives just for fun."[18] US Department of Energy secretary Jennifer Granholm agrees: "Me individually eating less meat is not going to do anything. And boy, wouldn't they love for us all to be distracted on our individual recycling plans. It is not what we need. We need big change, and that big change happens with policy. So, if anybody wants to do something on an individual level, *vote*."[19]

I recognize the ruinous behavior of industry as well as industry's and government's capacity to make needed systemic change. Yet this line of thinking leaves me and the millions of others concerned about the environment in a bind: If I do nothing, I feel worse than if I do something. I may become paralyzed by imbibing the poison of environmental despair. I can vote, but the next election is two years off. I want to do *something now* to address the environmental and climate crises along with the injustices and inequities that result.

Perhaps we are spending too much time on the individual-action-versus-industry-culpability-and-government-inaction argument. Egged on by the media and fossil fuel company disinformation campaigns, we are stoking acrimony among the very climate-concerned citizens with whom we could be working in harmony.[20] Worse yet, we are sowing doubt and contributing to the poison and paralysis of environmental despair.[21]

Why Take Individual Action?

As argued by the NASA climate scientist and activist Peter Kalmus, who has not flown since 2012 and converted his car to run on vegetable oil, trying to decrease the CO_2 molecules you personally churn out into the atmosphere is "less than 1 percent of the reason to take action."[22] So what are the other 99 percent of the reasons to take small individual actions such as avoiding flying or reducing food waste?

A Life Worth Living: Finding Purpose and Happiness

> **Global environmental change threatens the moral evaluation of our own lives, as well as of our generation, our communities, our nations, and humanity itself. This thought should be motivating. After all, who wants to be the scum of the earth?**
>
> —Steven Gardiner, philosopher, *Are We the Scum of the Earth?*

The first reason for taking small actions might be surprising: actions that reduce our CO_2 emissions can be healthy and rewarding, especially when performed with family and friends. Cooking a tasty vegan brunch, volunteering to transport uneaten food from a restaurant to a soup kitchen, or even participating in a challenge to raise money for a local climate nonprofit can add meaning to our lives and be an enjoyable way to spend time with friends, neighbors, and loved ones.[23] We may even acknowledge how minimal our power is to change the system but decide to use that minimal power nonetheless because it makes us feel purposeful and happier.[24]

Ikigai is the Japanese word for "purpose in life" or "life worth living." Older Japanese talk about ikigai as including activities such as "taking care of grandchildren," "volunteering," and "keeping their street clean and pretty." In one study of over forty-three thousand Japanese, having ikigai was linked to significantly lower risk of dying of cardiovascular disease, so much so that Japan's Ministry of Health, Labour and Welfare has made ikigai part of its official health-promotion strategy.[25] Similarly, taking climate actions can lend purpose to our lives. Given that climate change is already causing immense suffering, climate inaction may threaten our moral sense of who we are.[26]

Acting "morally" can even become an adventure, as Peter Kalmus writes in his book *Being the Change: How to Live Well and Spark a Climate Revolution*: "I've also reduced my personal CO_2 emissions from about twenty tonnes per year (near the US average) to under two tonnes per year. Overall, this hasn't been a sacrifice. It has made me happier. . . . When faced with some daily task—commuting to work, planning a trip, eating, showering, whatever—I began perceiving how it connects to our industrial system's preferred way of doing things, how it affects other beings and too often harms them. I began searching for alternative ways of doing things. This exploration often blossomed into adventure: unpredictable, fun, and satisfying."[27]

In a recent interview, Greta Thunberg reflected on her pathway from depression to youth climate activism. She talked about how the activists in far-off places have become her friends: "I know lots of people who have been depressed, and then they have joined the climate movement or Fridays for Future and have found a purpose in life and found friendship and a community that they are welcome in." Asked whether the best thing to come out of her activism was friendships, Thunberg responded, "Definitely. I am very happy now."[28]

Compared to making a big difference in lowering climate emissions, feeling good, having an adventure, or making friends may seem like trivial justifications for climate action. Who cares how we *feel* as the temperatures race to unprecedented heights? But people who are emotionally healthy are better able to act. For some, the climate crisis is leading to depression, paralysis, and despair. Taking action, especially with others, may counteract paralysis by instilling feelings of agency, purpose, hope, and connection. Taking action, then, becomes a form of coping, and hope and coping in turn are necessary for any action.[29] Furthermore, feeling good at having taken action spurs us to take additional climate action. This feedback between feelings and action is reinforced when we take action with friends and family.

Spillover

Could behaviors such as reducing meat consumption not only give us a purpose in life but also lead to other climate actions, including those aimed at policy change? In other words, is there a pathway for individual consumer actions, such as reducing food waste, to spillover to advocacy?

Sometimes people who engage in one climate-friendly behavior feel a moral license to indulge in other energy-intensive behaviors. A person might tell themself, "I am eating less meat so it's fine if I drive more. And since I am already eating less meat, why doesn't someone else pitch in?" This is referred to as a rebound or boomerang effect or "negative spillover." It turns out negative spillover is more likely to occur when people's climate-friendly behaviors are motivated by guilt, fear, or external rewards and when the effort required to perform the second behavior seems high. When those engaged in household behaviors believe that progress on climate is already being made without government intervention, they are also less likely to support government action.[30]

But when our decisions are based on our values (such as concern for other people or other life) and our identities (such as environmentalist, social justice advocate, or simply good neighbor), one climate action is more likely to lead to another. This is called "positive spillover." It can include not just one individual behavior leading to another, such as picking up litter leading to recycling, but also daily lifestyle behaviors leading to support for climate policies.[31] Taking a small climate action at home can help a person to develop a sense of responsibility vis-à-vis the climate, which makes them more supportive of carbon taxes and other climate policies.[32] Furthermore, when research subjects were prompted to reflect on how their sustainability behaviors were associated with their values, identity, and beliefs, they were more likely to support a carbon tax than those who weren't asked to reflect on their commitments. The researchers concluded that both individual actions to reduce greenhouse gas emissions *and* mulling over how those actions reflect one's fundamental beliefs are integral to individuals supporting the climate policies needed to transform energy and other systems.[33]

Fatima Delgado provides an example of spillover behaviors in Barcelona, Spain. While a graduate student, she regularly invited twelve friends to her apartment for sustainable tapas brunches. A year or so later, she wrote to me about how this initial behavior had spilled over into other activities. By then she had given a series of university and conference lectures about her Sustainable Tapas project. She had started marketing her ideas to local organic grocery companies. Although follow-up from the companies was slow, Delgado wrote to me in an email, "Nevertheless, I keep trying and looking [for] new collaborations and synergies across Europe. I truly believe in the project and believe it has the material to become a startup." Following up on the idea of a start-up, Delgado met with business innovation accelerators and entered her ideas into accelerator competitions. Those initial sustainable tapas brunches with twelve friends spilled over to university lectures, influencing businesses, launching a Sustainable Tapas social enterprise, and more recently, for Delgado, a position as a business professor mentoring students working with sustainable food start-ups.

Walking the Walk

Thus far, we have seen that meaningful climate action can make us feel happier and in some cases spill over to other consumer and advocacy actions. But there is another reason to take individual actions if we are aiming for systemic change brought about by climate policies. Simply put, people who live a climate-friendly lifestyle are more likely to have credibility when they advocate for climate policies. One study showed that community organizers who had solar panels on their own homes persuaded 63 percent more homeowners to install solar panels than community organizers whose homes lacked solar panels.[34] This is because those who actually engage in a consumer behavior are a more believable and authentic source of information about how to perform the behavior. Their actions also signal an urgency in responding to the climate crisis.

It turns out that there is a delicate balance, however, between being seen as a do-gooder who is doing everything to reduce their carbon footprint and ends up shaming others and a moderate who engages in a couple of actions and refrains from bragging about everything they do. It's admirable to be a vegan, buy renewable energy, weatherize your home, and avoid flying—but broadcasting all your virtuous behaviors may turn off your friends and neighbors. Evidence suggests that if you want to influence others, it might be better to admit that you are struggling and not doing everything rather than to be perceived as an annoying do-gooder.[35]

I once asked Brenda Mwangi (not her real name), a Kenyan professional who consulted on high-level climate agreements between the European Union and African countries, why she chose to eat a plant-rich diet. Given that she was working in high-stakes international negotiations, I thought a small action like changing what she ate would seem trivial. She told me, "When focusing on my professional work, which primarily focuses on climate change mitigation in lowering emissions, I don't want to just talk the talk—I want to walk the walk." Leading the way by acting in synch with your beliefs and recommendations can be an effective means of communicating the importance of climate action at multiple levels.[36]

Some skeptics might describe Mwangi as "performing" environmentalism when she dines on vegan cuisine with European policymakers and might deride such performances as useless. But in an article titled "All That Performative Environmentalism Adds Up: Don't Depersonalize Climate Change," *The Atlantic* writer Anne Lowrey argues otherwise: "The critics are right that focusing on individuals is a grave error if it obscures corporate culpability and systemic solutions. But I'm not about to get rid of my canvas bags and mason jars, buy a second car, or start taking short flights again. Talking with economists, climate scientists, and psychologists convinced me that depersonalizing climate change, such that the only answers are systemic, is a mistake of its own. It misses how social change is built on a foundation of individual practice."[37]

Lowrey doesn't want to get rid of her canvas tote bags because the bags personalize climate action. Such personalization creates a pathway from her individual action to systemic or policy change. But how?

Social Contagion: Influencing Neighbors, Family, and Friends

> **Overall, my network climate action was successful and extremely rewarding because I eventually had influence outside of my immediate network. Scaling up the project to include my housemates translated into them sharing plant-rich Bulgarian recipes with their families and friends.**
>
> —Siyana Popova, Cornell University student, Climate Solutions course

Greta Thunberg is perhaps the most famous of the climate activists who have taken individual action. She is vegan and does not fly. She even convinced her father to accompany her crossing the Atlantic Ocean in a sailboat, and her mother gave up an opera career to avoid air travel. Perhaps seeing her influence on her family and others, Thunberg recognizes her individual actions have more importance than simply what she herself does. She explains how through her actions she influences others: "It's true that if one person stops eating meat and dairy it doesn't make much difference. But it's not what one person does. This is about something much bigger. . . . We are social animals. We copy each other's behavior. I didn't stop flying or become vegan because I wanted to reduce my personal carbon footprint. It would be much more useful for me to fly around the world advocating for climate action. But it's all about sending a signal that we are in a crisis and that in a crisis you change behavior."[38]

I'm old enough to remember passengers smoking in planes and buses, guests smoking at dinner parties, and even being assaulted by my classmates' secondhand smoke when I ventured into the high school bathroom. Today my children complain about the faintest waft of smoke that reaches them when they walk by a neighbor who has been forced by his wife to smoke in the driveway. My how smoking norms—or what we deem as acceptable behaviors—have changed over the past fifty years!

Perhaps because we tend to follow the norms of the people we hang out with, smokers are more likely to have friends who smoke. In fact, smokers who manage to quit are not just doing something for their own health but also for the health of friends and family whom they influence to follow their example. This same "social contagion" occurs with environmental behaviors. If you were to fly low over neighborhoods in Arizona or California, you would observe that rooftop solar panels are not randomly distributed across the houses but rather are clustered. This is because neighbors—as well as family, friends, and colleagues—influence each other. If one person puts solar panels on their roof, their neighbors are more likely to follow. And if a person invites their housemates to eat healthy, plant-based meals, their housemates may be more likely to try plant-rich options in the future. Behaviors are subject to "social contagion"—they occur in spatial clusters such as neighborhoods and in clustered social networks of friends and family.[39] The words of social scientists Leor Hackel and Gregg Sparkman echo Thunberg's:

> We don't recommend taking personal actions like limiting plane rides, eating less meat, or investing in solar energy because all of these small tweaks will build up to enough carbon savings (though it could help). We do so because people taking action in their personal lives is actually

Cultivating Climate Behaviors in Others

You might not know it, but you may be what economists call a "moral rebel." Moral rebels are people whose strong ethics lead them to behave in ways others consider abnormal, such as eschewing beef, picking up litter, or spending every weekend canvasing for a climate-friendly mayoral candidate. The problem is that your behaviors may make others feel defensive about their own consumption habits or lack of political engagement. What can you do to prevent your friends from becoming defensive and even inspire them to change? Here are three communication strategies to consider:

- *Talk about actions, not character*. People will be more open to new behaviors than to changing who they are as a person. Recognize that many behaviors are habits and avoid judging climate-unfriendly behaviors as evidence of bad intent or poor moral character.
- *Talk about behaviors that you, and by implication others, can change*. Share a personal journey of how you changed a behavior, including any struggles along the way and ongoing challenges. Avoid using absolutist or morally laden labels such as "meat-eater" versus "vegan."
- *Commend any movement in the climate-friendly direction*. Give praise when someone tries out "Meatless Mondays" rather than condemning them for eating meat Tuesday through Sunday. Don't expect a goal to be reached all at once; it's a journey for all of us.[40]

one of the best ways to get to a society that implements the policy-level change that is truly needed. Research on social behavior suggests lifestyle change can build momentum for systemic change. Humans are social animals, and we use social cues to recognize emergencies. People don't spring into action just because they see smoke; they spring into action because they see others rushing in with water. The same principle applies to personal actions on climate change.[41]

Norms, Politicians, and Tipping Points

As climate-friendly behaviors spread among neighbors and friends, those behaviors become the local norm. Those not yet conforming to the norm jump on board. And as more and more individuals and neighborhoods demonstrate climate-friendly behaviors, they signal to policymakers that certain behaviors are becoming more acceptable. Importantly, these climate-friendly norms have just made the job of politicians and regulators easier. Now the politicians don't have to worry that supporting climate-friendly policies will backfire and lead to attacks from voters and other politicians. Rather, they feel they have the support of their constituents to take action and are more likely to propose new laws and regulations governing the climate.[42]

In a reinforcing circle, our individual actions influence social norms and in turn policies, which enable us to take additional climate-friendly actions.[43] Climate actions performed with others help forge the social connections, alongside the trust and shared identity, that become the foundation for future political action.[44] As the climate philosopher Dale Jamieson puts it, "individual action may not be very effective at directly reducing emissions but it can be very effective for consciousness-raising and enabling political action."[45]

Changes in behavioral norms also can become the basis for social movements. An example comes from changing norms around buying local foods, which has helped precipitate a sustainable eating or "vote with your fork" movement.[46] Norms about participation in climate activism are also changing among multiple demographic groups in the United States, especially the young. By 2019, 33 percent of African American teens, 30 percent of Hispanic teens, and 18 percent of white teens had participated in a climate protest, rally, or school walkout or contacted a government official to express their views on climate change.[47] Today we see the Extinction Rebellion, the Red Black and Green New Deal, the Climate Justice Alliance, the Climate Action Network, and numerous other groups coming together to form the climate justice movement.

As more and more people take action, we may even reach a social tipping point in that climate action has gained such widespread acceptance, it has become the societal norm.[48] Social tipping points are major transformations in norms and behaviors. Although they are triggered by a small change or intervention, they often follow a long period of slow buildup. One such small intervention was Greta Thunberg's Fridays for Future strike in 2019, which grew from a lone protester to somewhere near fifteen million global protesters in 217 countries over just two years. In Europe youth activists may be tipping public opinion, leading to election of green candidates and government climate action.[49] Another example is electric vehicles in Norway, where after a long phase of slowly increasing purchases, government incentives and accessible charging stations precipitated widespread buying of plug-in cars.[50]

When do individual actions to cut emissions tip the balance to broader systemic change? This happens when the direct actions are visible and thus have the potential to influence others. It happens when direct actions are part of a large collective effort, such as in the case of the Count Us In initiative calling for one billion actions to reduce emissions.[51] More broadly, it happens when individual actions, such as eating vegan, signal to government a willingness to accept regulations or to business an interest in buying plant-based foods. And as such foods become available in grocery stores, school lunches, and hospital meals, our direct lifestyle actions will have created new conditions that enable those healthy eating behaviors to spread. Finally, it happens when some individuals go beyond lifestyle changes to join organizations that advocate for climate justice. Thus, individual acts to cut emissions are important because they can lead to collective actions that promote wider and systemic change.[52]

An Ecosystem of Transformation

We often pit direct consumer action against indirect actions such as advocacy and protest. But like much in life, these distinctions are not always clear-cut. I began my climate action by limiting meat and dairy consumption, reducing food waste, growing my own food, using personal heating devices so I wouldn't have to heat the whole house, and walking to work. Although not directly advocating for different policies, I was part of the ethical consumption movement.[53] Then, wanting to do more, I joined Elders Climate Action and the Climate Reality Project and downloaded the Climate Action Now app, which facilitates easy advocacy actions. Now I am working with these organizations to help members explore plant-rich eating and to advocate for healthier and climate-friendly food and farm policies.

In the end, whether you eat less meat or join in a protest, you are still one individual. In a sense, all actions are individual actions. Jamieson recognizes that:

"When it comes to voting, writing letters, making modest campaign contributions, or even occupying Wall Street, it is hard to feel that my individual act has much efficacy."[54] But we can try to spread our individual actions to our family and friends to scale up our impact, and we can join climate nonprofit organizations to escape feeling that our individual voting and letter writing are futile. Acting collectively with policy-savvy organizations enables us to effectively influence government and business policies.[55] And climate and social justice organizations are forming networks themselves, which have become the backbone of the climate justice movement.[56]

Perhaps it's time to lay down the battle-axe. Time to recognize how those of us who are changing daily behaviors, such as what we eat, can work together with those of us focused on changing policy through protests and joining climate organizations. Time to realize that the same person often engages in both types of actions. And time to see how all these actions are connected and can reinforce each other.

As the climate scientist Michael Mann has pointed out, the debate pitting consumer behaviors against political action has driven an unnecessary wedge between proponents of one type of action and another.[57] Yet we can extract this wedge if we embrace the facts. First, consumers are responsible for a sizeable portion of our greenhouse gas emissions—more than 60 percent.[58] Second, government policies, such as rebates for installing solar panels, can incentivize climate behaviors. But here's the catch. Government policies are more likely to be instituted and accepted if a grassroots effort has already resulted in a sizable portion of citizens engaging in the behavior. This is because individuals demonstrate the feasibility of these actions and shift people's perception of what is accepted or normal behavior.[59] Feedbacks between individual behavior and shifting social norms, alongside government as well as nonprofit and business interventions, constitute a "broad *ecosystem of transformation*."[60]

In a nutshell, taking climate action is about our own well-being. It is about paving the way for new climate actions and for new people to join our actions. And it is about laying the groundwork for transforming societal norms and government policies.

Choosing Effective Climate Actions

Paul Hawken reminds me of the Tom Hanks movie character Forrest Gump, who just happened to be present for every important historical event of his time. From the civil rights movement, where he was assaulted by the Ku Klux Klan while photodocumenting voter registration drives in Bogalusa, Louisiana,

Hawken moved to producing concerts for the likes of Janis Joplin and the Grateful Dead, to converting a basement co-op into a multi-million-dollar organic food company, and then launching an upscale garden supply company for yuppies. One of Hawken's more recent endeavors demonstrates once again that he is capturing the moment. In 2014 he launched Project Drawdown, which has had a major influence on how we think about effective climate solutions.[61]

The idea of Project Drawdown—that is, compiling a list of climate solutions or actions humans can take to draw down greenhouse gases in the atmosphere—emerged in the late 2000s when Professor Amanda Ravenhill invited Hawken to speak to her students at Presidio Graduate School in San Francisco. Ravenhill's students went on to conduct the preliminary research for Project Drawdown. Two things made Hawken's idea of "drawdown"—the point where we start to reduce rather than increase atmospheric carbon—unique. First, the focus was on solutions—what we want for the future—whereas the vast majority of climate news stories at the time were about the future we didn't want. Second, it included strategies not just to lower emissions but also to draw down carbon already in the air, through such actions as planting and caring for trees, restoring abandoned farmland, or rewetting peatlands.[62]

Hawken then assembled hundreds of scientists to conduct a rigorous review and assessment, resulting in a list of eighty-two climate solutions ranked by their effectiveness in drawing down greenhouse gases.[63] Project Drawdown's ranked list of solutions is invaluable because we too often focus on less effective solutions such as recycling, perhaps because the fossil fuel giants have deluded us into thinking that by simply tossing things into different bins we can be virtuous environmental citizens and solve the environmental crisis.[64] In reality, most recycled plastics, paper, glass, and aluminum end up in landfills due to lack of facilities to process waste and inability to refashion waste into something useful. In fact, recycling ranks only forty-second on Project Drawdown's list of solutions.

For the millions of people around the globe who witness the climate changing and want to do something about it,[65] Project Drawdown helps us answer the question, What are the most effective actions I can take to address the climate crisis? Of course, the answer will vary according to how much time and money you have as well as where you live and the opportunities for accessing particular climate actions. A low-income resident in the United States may not be able to access public transportation or readily drive to stores that sell healthy, plant-based foods, while a rural resident in Zimbabwe may need to eat locally available chicken or goat in order to get enough protein. Of course, lots of other factors impact our climate actions—what we see others doing, what actions signal a valued identity such as environmentalist or good neighbor, and what is easy to implement. Unfortunately, companies mounting corporate sustainability campaigns while lobbying against climate legislation, or the governors of meat-producing states belittling efforts of neighboring states to reduce meat consumption, can lead astray people trying to be good climate citizens.[66]

Regardless of what influences you, you can still be informed of the most effective actions you can take, given where you live, your resources, whether you have children in your household, and your other circumstances.[67] As an example, for those with access to grains, legumes, nuts, and vegetables, switching to a plant-rich diet is four times more effective at reducing emissions than switching to LED lightbulbs and nearly twelve times more effective than recycling. Many do not realize that reducing food waste ranks even higher—over five times more effective than LED lightbulbs and sixteen times more impactful than recycling—in reducing emissions.[68] One high-ranking solution, "family planning and education," captures the "cascading benefits of access to universal education and voluntary family planning." In carefully worded language, it avoids any suggestion of infringing on reproductive autonomy, which has a fraught history, given attempts by richer countries to curb population growth in the developing world. [69]

If you read the figures on how different actions impact greenhouse gas emissions, you'll realize quickly that Project Drawdown's exact numbers are not necessarily reproduced in other publications. This is in part because some calculations take into account global emissions and some limit themselves to one country. Some organizations' calculations only consider carbon sources (emissions), whereas others include both carbon sources and sinks (actions such as reforestation that remove carbon from the atmosphere). Complicating the situation further are the facts that solutions take different amounts of time to have an impact on greenhouse gas levels and solutions also interact with and affect each other.[70] How do you compare the impact of the onetime purchase of an electric vehicle versus composting food wastes daily for years or donating to a climate justice nonprofit? How do you accurately account for the fact that putting solar panels on your roof might influence your neighbors to do the same and eventually change the norms about how we meet our energy needs, not to mention lower the price of solar power? Recognizing that no way of calculating emissions is perfect, many people agree that Project Drawdown's list of global climate solutions is based on solid science and is an easy-to-use guide to effective actions.

So, what solutions rank the highest? Among the top five most effective Project Drawdown solutions are "plant-rich diets" and "reduced food waste"—or, alternatively, eating healthy meals featuring fruits, vegetables, grains, nuts, and legumes and being creative with leftovers. These two highly ranked solutions are actions we can do in our everyday life, so-called lifestyle or consumer behaviors or direct actions to lower our emissions.[71]

Of course, not every climate solution is something we can incorporate into our daily routines. Project Drawdown solutions such as "refrigerant management," "onshore wind turbines," "tropical forest restoration," or "improved cook stoves" lend themselves to regulation, legislation, or in some cases philanthropy. Here, contributing money to or volunteering with civil society organizations knowledgeable about influencing climate policy through legislation, regulations, protests, lawsuits, boycotts, voter registration drives, and online letter-writing parties are ways to contribute to our collective project of drawing down greenhouse gases. Rather than directly reduce emissions, these collective advocacy actions draw down greenhouse gases indirectly by spurring new policies.[72]

Choosing Effective Climate Actions

Choosing a climate action depends in part on what makes sense for you, given your interests, abilities, and constraints. You might also want to consider how an action scores on three other criteria: technical potential, behavioral plasticity, and the likelihood of government taking action.[73]

Technical potential is the reduction in emissions if the action were universally adopted. One high technical potential behavior is a vegan diet, which can cut our global food-related emissions by 49 percent. If adopted in the United States, where per capita meat consumption is three times the global average, dietary change could reduce food emissions by up to 73 percent.[74] Another high technical potential behavior is low-carbon transportation such as biking or taking the bus. Nearly 30 percent of emissions in the US and Europe stems from transportation.[75]

Behavioral plasticity refers to the likelihood we can change the behavior by an intervention, such as inviting friends to participate in the behavior alongside us. Multiple factors influence behavioral plasticity, including whether a behavior is performed frequently, whether it is a habit, what social norms surround the behavior, and whether the behavior is part of a cultural tradition such as eating meat to celebrate family events.[76] In addition to qualities of the behavior itself, we need to consider characteristics of the people we are trying to influence, such as their belief in their ability to change a behavior, their income, their available time,[77] and their access to alternatives such as plant-rich foods or public transportation.

Likelihood of government action: A final strategy in choosing a climate action is whether government is likely to step in and regulate the behavior. As of 2022 the US government appears willing to regulate car emissions and promote electric vehicles. But one small attempt by Governor Jared Polis of Colorado to promote going meat-free one day a week backfired, illustrating how difficult even talking about cutting down on meat is in the United States. Upon assembling a team of agriculture bureaucrats and industry officials, Governor Pete Ricketts of Nebraska, a major meat-producing state bordering Colorado, declared that his neighbor's action was a "direct attack on our way of life." He then proclaimed Saturday "Meat on the Menu Day" in Nebraska.[78] Not to be outdone, Iowa governor Kim Reynolds declared April "Meat-on-the-Table Month." And Iowa senator Joni Ernst introduced the TASTEE Act—Telling Agencies to Stop Tweaking What Employees Eat—which would prohibit federal agencies from establishing policies that ban serving meat to employees.[79]

Furthermore, a study of dietary guidelines from eighty-five countries revealed that they rarely consider greenhouse gas emissions. In fact, if we all followed the North American dietary guidelines, we would exceed by 300 percent the emissions that would be needed to reach the Paris Agreement of limiting warming to less than two degrees Celsius.[80]

Climate Justice

Perhaps given Paul Hawken's early involvement in the civil rights movement, it's surprising Project Drawdown did not originally focus more on climate *justice*. Then again, in the past few years, concerns about justice, spurred in part by the many people of color leading the climate justice movement and the murder of George Floyd, have catapulted to the forefront of the US and global climate movement. (In fact, in his latest Forrest Gumpian accomplishment, Hawken came out with a new book in 2021, *Regeneration: Ending the Climate Crisis in*

One Generation, that weaves together justice, equity, and human dignity with climate and biodiversity.)

Organizations such as the Climate Justice Alliance, the Red Black and Green New Deal, Interfaith Power and Light, and Coalfields Development demonstrate that climate is not just about science—it's also about justice. The United States and global communities recognize we *cannot*, in good faith, solve the climate crisis on the backs of those who have suffered most from the age of fossil fuels. Coal miners allowed us to get rich off coal power but suffered from black lung disease, poverty, and the opioid epidemic.[81] So how can we bring economic and racial justice to coal mining communities—and to thousands of other communities impacted by fossil fuel pollution—as we transition away from coal, gas, and oil? More narrowly focusing on the high-ranked Project Drawdown solution of plant-rich diets, how can we make non-animal-based protein and fresh foods available and affordable for rural residents who currently have to drive miles to find a place to buy food—only to arrive at a gas station convenience store where the best alternative to meat is row upon tempting row of chips and candy? Or to impoverished communities living next to polluting factories and highways in cities that have been labeled food deserts due to the dearth of grocery stores? And how do we help the farmers who for generations have raised cattle or the immigrants laboring in meat-packing plants trying to make enough money to pull themselves out of poverty? Or account for those living in poor countries where livestock is the most available source of protein? A climate justice perspective asks us to look at the energy, food, and transportation transitions needed to avert the worst of the climate crisis through the lens of not only environmental issues but also issues of social and economic equity.[82]

Yet in places with access to sufficient nonanimal protein, a plant-rich diet is also a healthy diet, including for the millions of people who suffer from obesity and heart disease in the United States and a growing number of other countries. Similarly, another top Project Drawdown solution, reducing food waste, saves money and thus benefits all families, especially those with lower incomes. But just as our ability to eat a plant-rich diet is determined in part by our access to nonanimal foods, our ability to divert food from the waste stream is partly determined by whether we have reliable refrigeration or whether our city has a food-donation nonprofit or a curbside food-waste collection program that is affordable and convenient.

For some people, the highly ranked Project Drawdown solution "health and education" raises a huge climate justice red flag.[83] It seemingly attempts to capture the impact of population growth on greenhouse gas emissions.[84] As women gain education, they have fewer children, which in turn means fewer people driving, flying, eating meat, heating homes, and otherwise emitting greenhouse gases.

The problem is that such initiatives are largely targeted at poor communities in developing countries, whereas the rich person's average climate emissions can be over one thousand times that of an individual living in a poor country. Thus, wouldn't it make more sense to reduce population growth in wealthier countries?[85] The problem is compounded by centuries of rich countries attempting to control the population of poor countries and slave owners and later eugenicists and white supremacists trying to control the populations of black and brown and disabled people. These population-control campaigns used horrific methods, such as forced sterilization, which became so commonplace as to garner the colloquial name "Mississippi appendectomy."[86]

Likely in response to these and other equity concerns, the project Drawdown Lift was launched in January 2021 to connect issues of health, education, poverty, and climate in poor countries. According to Drawdown Lift, "when we work together to address societal inequities by lifting up gender equality, universal education, and sexual and reproductive health and rights, we can also advance long-term solutions to climate change."[87]

I would be remiss in talking about inequity and injustice without recognizing that I, along with many readers of this book, are among the global rich. From 1990 to 2015, a period when cumulative emissions doubled, the richest 10 percent of the world's population, or about 630 million people, was responsible for 52 percent of the cumulative carbon emissions. In contrast, the poorest 50 percent, or about 3.1 billion people, was responsible for just 7 percent of cumulative emissions.[88] So, people like me may not have the same culpability as the fossil fuel barons, but we are all part of systems of injustice—and may even want to do something about it.

This Book: Action!

People can act directly to reduce climate emissions in their daily lives and can act indirectly to pressure government and business to enact practices and policies that reduce greenhouse gases.[89] Of the four actions highlighted in this book, two directly reduce our own emissions—*eating less meat and dairy* (a plant-rich diet) and *reducing food waste*—and two are indirect actions—*donating* to and *volunteering* with organizations whose work fosters climate justice.

Why does this book focus on these particular actions? Eating a plant-rich diet and reducing food waste are things we can do in our daily lives that can be fun, healthy, save money, and make us happy, while providing opportunities to spend time with family and friends. They are actions that scientists have shown will make a big difference in drawing down greenhouse gases. We ponder what food to eat multiple times every day. This frequent decision-making can, in turn, lead to reflection about our responsibility for reducing emissions and support for other climate actions, including those that involve policy change.[90] And adopting a plant-rich diet, in particular, is an action that the government is less likely to create policy around, compared to other important sectors such as transportation and energy or even food waste. Although using carbon-friendly transportation and energy are super important, we generally have less control over them relative to choices about food and food waste.[91] Nevertheless, we can address these issues by supporting climate organizations that advocate for policy change, through donations and volunteering. Thus, this book focuses on lifestyle changes we can do in our daily lives and on influencing policies and practices around climate solutions that are necessarily addressed through laws, incentives, and regulations.

I have tried to choose actions that are accessible to many people. Although eating nuts, grains, and legumes is generally cheaper than eating meat,[92] plant-rich foods are not accessible to those living in food deserts and not advisable for those whose only source of protein is meat or dairy. But for most of us, choosing to eat a certain way or choosing to reduce our food waste is more doable than other critical climate actions such as biking to work or buying a new induction stove. Similarly, many people may not have the extra funds to contribute money to a nonprofit group (although statistics show that the poor and rich give a similar proportion of their income to charity[93]) or the extra time to volunteer for a climate organization. In addition to access, I have tried to be sensitive to social equity, especially because our vast social inequalities contribute to environmental degradation.[94] And while recognizing the importance of transformational movements such as degrowth or even more radical actions such as physically blocking access to or even damaging pipelines,[95] in this book I do not fundamentally challenge our existing economic and political systems. Rather, I aim to help as many networks of family and friends as possible to engage in meaningful climate action, here and now.

Understanding the latest research and "theories of change"—that is, assumptions or hypotheses about which interventions lead to specific behaviors and to social transformations—helps us to reflect on how we are trying to influence people and how we can be more effective. Thus, in each of the first four chapters in this book, I introduce research reflecting a different theoretical lens that explains how actions can have impacts beyond individuals. Although I link specific theories to specific behaviors—social contagion theory for plant-rich diet,[96] practice theory for food-waste reduction,[97] public-goods theory and game theory for philanthropy,[98] and efficacy, organizational, and social movement theories for volunteering[99]—the theories and research in one chapter often can be applied more broadly to understanding behaviors discussed in other chapters. For example, our behaviors are influenced by what we see our housemates doing. But housemates may be more likely to influence what we eat than to influence which organizations we donate to. This is because we eat together, whereas donating is often a private behavior. Thus, research on how behaviors spread in social networks is the focus of chapter 1 on plant-rich diet. However, as the rise of social media transforms donating from writing a check alone in a study to joining an online GoFundMe campaign or Ice Bucket Challenge, the potential for influencing family and friends increases. This means that the social influence theories covered in chapter 1 on a plant-rich diet are also relevant to chapter 3 on donating. By explaining the research and theories on how to spread behaviors that are effective in drawing down greenhouse gases, my goal is to empower readers to not just take climate action themselves but also to involve their family and friends in taking that action alongside them—that is, in "network climate action."

The final chapter of this book revisits the question of individual and corporate responsibility, this time from the perspective of philosophy. I share how the work of two feminist philosophers, Iris Marion Young and Robin Zheng,[100] has helped me to sort out questions of where my responsibility lies in the struggle for climate justice and hopefully can help the reader do the same.

In addition to sharing my stories and elucidating social sciences research, this book recounts the tales of climate activists and students I have encountered through my teaching. The activists, whom I refer to as fellows and who hail from countries around the world, participated in the Cornell Climate Online Fellowship, which entailed a series of lectures, readings, and discussions, followed by taking a network climate action. You have already met two of them—Fatima Delgado from Ecuador, who launched Sustainable Tapas in Spain as her Cornell fellowship project, and Brenda Mwangi, the Kenyan climate consultant who talked about how she walks the walk by eating plant-rich meals when meeting with European Union officials.[101] In addition to the fellows, I recount anecdotes about participants in a global online Network Climate Action course as well as students who participated in a Cornell undergraduate Climate Solutions course. Like the fellows, each student completed a network climate action with family and friends. In the end, descriptions of these climate-concerned citizens of the United States, Kenya, Germany, Zimbabwe, India, and other countries, coupled with digests of research on how we influence each other and the musings of environmental philosophers, reveal that all of us are responsible for the climate crisis, albeit to different degrees.

At the heart of this book are connections. These are the connections we make with family and friends when eating climate-friendly meals together—connections that foster trust and shared norms and that become the yeast that bubbles up to collective action.[102] These are the connections between the actions

1

EAT

Before COVID-19 and before I found it harder and harder to justify air travel, I used to fly to China for my work in environmental education. My friend and former PhD student Yue Li and I led workshops at Hangzhou Botanical Gardens, at a school in Hong Kong, and at China's Nature Education Society conference in Wuhan. We met with Chinese urban sustainability advocates on a rooftop garden in Shenzhen, with the Alibaba Foundation at its lush headquarters in Hangzhou, and with the Ministry of the Environment in Beijing. And we took side trips to hike in the Yellow Mountains, to visit Yue's family in Yunnan, to walk up and down the golden rice paddy-strewn hills and raft past the mysterious cliffs of Guilin, and to join her parents on a river cruise up the Yangtze River from the Three Gorges Dam.

Being a local, Yue always chose the restaurants where we dined. During our last visit to Beijing, knowing that I prefer to avoid eating meat, she chose vegan options. My impression is that a plant-rich diet comes naturally to many Chinese, given that each restaurant or university canteen where we ate presented a panoply of novel greens and vegetables. There often were small bits of pork and chicken mixed in with the vegetables, but I didn't find large slabs of beef or even chicken thighs as the main dish. And there were the occasional strictly vegan restaurants, which tended to come in two flavors: those that celebrate vegetables and offer colorful dishes with intriguing flavors and spices and those that try to mimic flesh-based meals. At one imitation-style establishment, the waiters delivered to our table a sizzling "fish" peering out from its aluminum foil wrapper, rendered in soy protein and faithful to the last detail—eyeballs, gills, and all.

At every meal during my stays, we were accompanied by friends, family, and colleagues: dinner with university students, lunch with Yue's thirty-something college buddies, a feast with Yue's extended family at the faux-fish vegan restaurant.

I appreciated Yue seeking out plant-based restaurants for these group occasions, both to accommodate me and because she understands that engrained behaviors, such as what we eat, can be changed through the influence of family and friends. But is it reasonable to assume that Yue's family, friend, and colleague networks—or any other network for that matter—will take up veganism after a single meal together? How exactly do friends and relatives influence each other's eating habits?

Eating is a complex behavior. It's influenced by taste and tradition, what's available and what's habitual, and even a desire to feel close to a loved one who has died.[1] Eating is also shaped by whether we have access to healthy foods and what we have learned to cook. This complexity means eating habits are stubborn and resistant to change.

But people do occasionally manage to change complex behaviors and can even spread new behaviors to family and friends. Because what we eat impacts our climate so significantly, this chapter is about food choices and how, through network climate action, we can spread our behaviors to our own friends and family and possibly even beyond.

Why Go Vegan, Flexitarian, or Climate Carnivore?

> **[Without considering] the health of our planet, I think it's really futile to talk about the health of the human population. They are all interrelated and intertwined, and they have to be considered and looked at simultaneously rather than separately.**
>
> —Frank Hu, nutritionist, quoted in Sweet, "Health Plan, Healthy Planet"

According to Project Drawdown, reducing meat consumption, along with reducing food waste (see chapter 2), go much further than other lifestyle or direct actions, such as installing LED lightbulbs, to draw down greenhouse gases.[2] Food production contributes up to 30 percent of total greenhouse gas emissions. (It also accounts for 70 percent of freshwater use and occupies nearly 40 percent of our planet's land.[3]) The stark reality is that humanity will not manage to limit global warming to 1.5 or even 2 degrees Celsius without tackling food systems.[4]

Our individual and social eating habits play a big role in reducing greenhouse gas emissions.

If everyone in the world followed a healthy diet with no more than three portions of red meat per week, by 2050 our food-related greenhouse gas emissions would decline from nearly 11.5 to just over 8 gigatons of CO_2 equivalents per year. If we all went vegetarian, food-related greenhouse gas emissions would decline to 4.2 gigatons CO_2 equivalent per year, whereas if everyone on the planet went vegan, the emissions number would be 3.4. This would result in a whopping 63–70 percent reduction in food-related emissions by mid-century compared to our current trajectory.[5]

Measuring Gases in the Atmosphere

A gigaton is one billion tons. In 2019 humanity released about 48 gigatons of CO_2 equivalent, more than a quarter of which came from food systems. Scientists estimate that under the current Paris Agreement, global emissions will reach about 55 gigatons CO_2 equivalents per year. Reductions beyond those agreed upon in Paris are needed to keep global temperature increases below 2 degrees Celsius.[6]

Even if a person chooses not to become a vegan or vegetarian, substituting fish, chicken, and eggs for red meat and cheese and eating more plant-based meals made from beans, wheat, rice, nuts, and the growing number of meat substitutes can still make a significant difference. You might choose to become a flexitarian who tries to reduce meat and dairy as much as possible but who doesn't feel guilty (or abstain from) partaking of a roast pig or turkey at a family celebration or having a hamburger hot off the grill during a graduation party. If all of us adopted a flexitarian diet, we would reduce CO_2 equivalents by about 5 gigatons per year—not that different from the vegetarian diet. Or you might prefer to become a climate carnivore, someone who replaces 75 percent of beef, lamb, and dairy with pork and chicken. If everyone became a climate carnivore, we would still reduce CO_2 equivalents by over 3 gigatons per year.[7] In fact, Americans already reduced per capita beef consumption by 19 percent from 2005 to 2014 and in the process avoided emitting approximately 0.27 gigatons of CO_2 equivalents. This represents a 10 percent reduction in greenhouse gas emissions from foods[8]—a trend that needs to continue if we are to reach the point when we are drawing down greenhouse gases in our atmosphere.

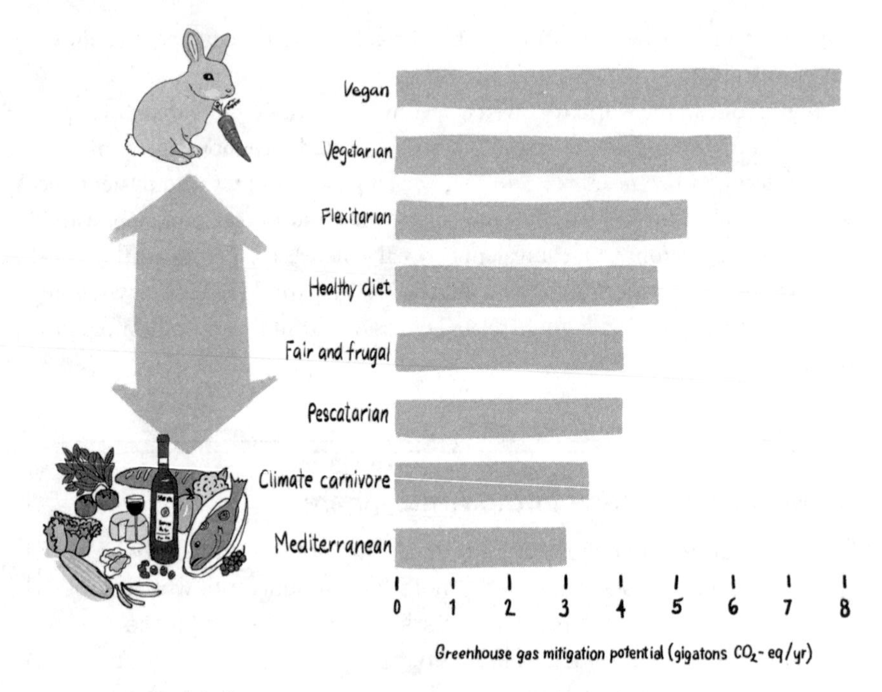

Amount of CO_2 not emitted globally by 2050 were people to adopt different diets. Chart by Emily Hopkins, redrawn data from Mbow et al., "Food Security," 2019.

The worst meat to eat, in terms of climate and the environment, is beef. Not only cattle but also other ruminants such as sheep and goats burp large amounts of methane into the atmosphere. Because cheese and butter are produced by cows and other ruminants, these dairy foods also have high greenhouse emissions. The culprit is the ruminant's digestive system—its four-chambered stomach churns the corn, grass, and whatever else it eats. As bacteria, fungi, and other microbes inside the cow help digest its food, they also produce methane—a greenhouse gas eighty-four times as potent as CO_2 over twenty years.[9] When cattle burp, the methane exits the stomach and enters the atmosphere. Because people raise lots of cattle (the world cattle population is over a billion[10]) and because cattle burp frequently, they are a whopper when it comes to climate change. In the United States, beef production accounts for over a third of total greenhouse gas emissions associated with food production.[11]

Consuming pigs, chicken, and fish—none of which belch large quantities of methane—results in about one fifth to one seventh of the greenhouse gases emitted by eating beef. Production of butter and cheese, foods common in vegetarian diets, emits more methane than producing pork, chicken, eggs, or the new plant-based meats such as Beyond Burgers. Nuts, soybeans, and wheat are way down there on the emissions scale, whereas rice production is five times

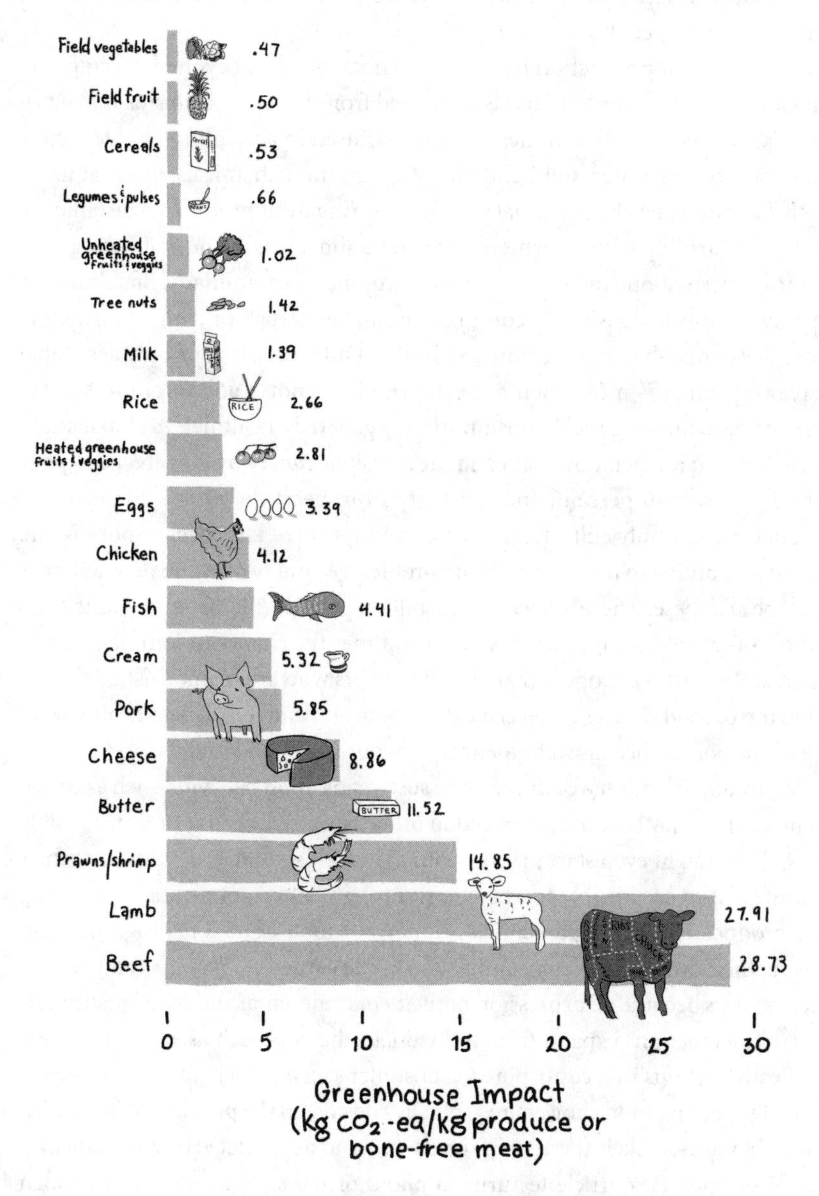

Carbon emissions of different foods. Chart by Emily Hopkins, redrawn data from Clune, Crossin, and Verghese, “Systematic Review of Greenhouse Gas Emissions for Different Fresh Food Categories,” 2017.

higher than wheat (but still way below beef).[12] Here again methane extends its smelly tentacles: rice is often grown in flooded paddies where anaerobic bacteria emit this potent greenhouse gas.

Lowering meat production has environmental benefits beyond lowering carbon emissions. It means less land is converted from forest to the farms that grow the crops to feed livestock. It means less water used in agriculture and less contamination of our water, soils, and air by agricultural chemicals and manure.[13] For those concerned about animal welfare, lowering meat production also means less factory farming with its gruesome toll on animal health and well-being.

But concern about the climate, the environment, or animal welfare are not the only reasons to consider becoming a vegan, flexitarian, or climate carnivore. Those diets are also better for humans' health. Obviously, if your only accessible source of protein is meat, then a meat-free diet is not a good option, but for everyone else, limiting meat consumption is generally healthier. A global study concluded that a vegetarian diet can reduce risk of cancer by 10 percent, type II diabetes by over 40 percent, and mortality from heart disease by 20 percent.[14] Although not the only culprit, rampant consumption of fatty beef or pork is one factor contributing to the global obesity problem. About two billion people across the globe are obese, whereas fewer than a billion people suffer from malnutrition. This means a child born today has over two times the chance of joining a family with members who are obese than with members who are malnourished.[15]

On top of food choices being critical to our health and the planet's health, what we eat is important because it influences what our family and friends eat. And once our family and friends have adopted one sustainable food behavior, such as eating less meat, they may become interested in other aspects of healthy and sustainable eating. They might even start to "vote with their fork"—that is, demonstrate their commitment to sustainable food systems through behaviors such as purchasing local produce and fair trade coffee, supporting food-donation programs, and growing their own food in community and home gardens.[16] People's food-related practices thus become an expression of their environmental and social justice values, and their actions expand from individual behaviors such as eating less meat to collective actions that contribute to consumer social movements.

Finally, because food choices are generally considered the purview of individual choice, they are less likely than energy or transport to be regulated by government.[17] In a *Washington Post* article featuring a photo of Barack Obama and Joe Biden chowing down on hamburgers, US secretary of agriculture Tom Vilsack responded adamantly to Republican fearmongering about the government coming after our meat: "There's no desire, no effort, no press release, no policy paper—none of that—that would support the notion that the Biden administration is going to suggest that people eat less meat. . . . Or that U.S.D.A. has some program designed to reduce meat consumption. It's simply not the case."[18] Clearly, the government is reticent to regulate or even incentivize plant-rich diets. This reluctance throws food production remedies back into the arena of individual and social behaviors. If enough people show their willingness to turn away from meat, government may eventually respond with policies that support plant-rich eating. In the meantime, to influence what people choose to eat, we first need to understand factors that determine our complex eating behaviors.

Why Am I Eating This?

Recently a Russian exchange student we had hosted nine years back paid us a visit while in the United States to serve as a bridesmaid in her classmate's wedding. We planned one of our typical family feasts to welcome Adel Davletgaraeva. I made blueberry pie with a butter-pastry crust. I knew I could substitute a canola-oil crust, but over many years my flaky buttery crusts have become a favorite with my family. I also made squash-corn bread with squash and jalapeno peppers from our garden. I mixed in grated cheese because we had rescued some slightly moldy cheese from my mother-in-law's overstocked refrigerator, and my son lobbied for adding it. My husband cooked up venison with garlic and onions. Deer have four-chambered stomachs like cattle and are high methane emitters; they are also so abundant in upstate New York that forests are no longer regenerating, and it's practically impossible to grow anything in our downtown Ithaca garden without constantly chasing off the deer. My husband hunts deer during the local hunting season.

Recreating Family Traditions, Creating Community

Regardless of whether my family's meal to welcome back Adel was sustainable or not, the decisions that went into our dinner illustrate that eating practices are notoriously hard to change. In a 2020 opinion piece in the *New York Times* titled "I Admire Vegetarians. It's a Choice I Won't Ever Make," the Chinese-Malaysian American columnist Alicia P. Q. Wittmeyer recounts the tale of partaking in a whole pig roast to celebrate her mother completing chemotherapy.[19] Wittmeyer describes how she values her Malaysian cooking traditions every day and how she will never give up her Malaysian chicken rice dish. She feels a twinge of guilt at indulging in pork and chicken given how dire the climate crisis is. But as she watches her friends one by one become vegans or vegetarians, she decides that, for her, holding on to family and cultural eating traditions is paramount.

Despite Wittmeyer's feelings of guilt, her eating practices are actually making a contribution to mitigating climate change, at least compared to if she were eating beef, lamb, or goat. Judging by her description of her chicken and pork dishes, she might be a climate carnivore who avoids beef and other ruminant meats and dairy. If you are like Wittmeyer, trying to hold on to family traditions while being painfully aware of the environmental (and health) crises promulgated by profligate meat consumption, don't despair or feel too guilty about cooking a family chicken dish or attending a celebratory family pig roast. You may be contributing more to reducing greenhouse gases than you realize. Perhaps you can do a little bit more. Cook with chicken more often than with pork. Pursue a flexitarian diet rich in fruits and vegetables, nuts, grains, and beans. And hold off on steaks and hamburgers. The planet may thrive (and remain a home for us) longer because of these efforts and because networks of family, friends, and colleagues follow our lead.

Also, consider additional climate-friendly ways to gather with families, neighbors, and communities. In the 1990s I initiated the Garden Mosaics program to acquaint young people with the diversity of cultures and plants present in urban community gardens. As I visited community gardens in New York, Houston, Minneapolis, Sacramento, Toronto, and Soweto, South Africa, I learned how African Americans who migrated from southern farms to northern cities, Korean grandparents who left their country to be with their children and grandchildren but were left alone during the day, Laotian and Afghan refugees seeking a place that offered solace, Lower East Side folk artists needing a venue to express themselves, and township dwellers trying to build a new, post-apartheid future all came together to grow food and community in gardens. The community gardens also provided a site for concerts, activism, and picking up produce from farmers delivering to subscribers of a CSA. ("CSA" refers to community-supported agriculture, an arrangement whereby consumers purchase directly from farmers, often through a weekly farm produce delivery during the growing season.) I was experiencing the urban version of what my Cornell colleague called "civic agriculture," or alternative food systems that "reconnect farm, food, and community."[20] Sharing harvests from a community garden can bring together people just as well as barbecues or bowling leagues.[21]

Our food choices are deeply telling. What we eat reflects our obligations, sacrifices, responsibilities, and even love.[22] We learn to eat and prepare food from those we are most intimate with, family and friends. Interestingly, what we eat is so embedded in everyday activities with others that when asked to justify eating behaviors, people often talk about being a parent, a spouse, or a friend.[23] But it's not just social influences that determine what we eat.

"Plant-rich doesn't mean houseplants, Jason."

Old Habits

When you're in your comfort zone, you just go with what's in your mind and eat what you're used to.

—University student, quoted in Gallimore, *Understanding the Reasons for and Barriers to Becoming Vegetarian*

My friend Brian Reed (not his real name) was at the wheel as we drove along the interstate from Ohio through Pennsylvania and into New York. The miles upon miles of cornfields were mind-numbing, giving us plenty of time to talk—about corn, climate, and the food-industrial complex. My friend, who happens to be vegan, was getting hungry. I offered him some healthy nuts, but he waved them away. Not to worry, he said, as he pulled into McDonald's and ordered a bag of greasy, salty French fries.

We can't change the fact that many people, myself included, love the taste of salt and grease. Because eating a hamburger and French fries offers us pleasure or because we grew up eating burgers and fries or because hamburgers and fries are easy to find along the highway, eating a hamburger and fries can become a habit whenever we travel and whenever we go into a fast-food restaurant.

Habits—that is, behaviors that we repeat frequently and without much thought—account for nearly a fifth of our eating behaviors.[24] Habits are triggered by situational cues. In the morning, as soon as I walk into my kitchen, I take the same jar of granola from the shelf and jar of berry sauce from the refrigerator. Although I could grab something else for breakfast, it might be hard to break my habit of eating the same type of cereal every morning. The situational cue—my kitchen at seven o'clock—and the muscle memory born of repetition trigger my behavior.

Because habits are performed without thinking or even our awareness that we are doing them, information is not part of our decision-making calculus. If the government adds a label on my cereal box telling me the sugar content of my granola—an intervention intended to reduce diabetes and obesity—I'll likely not even notice because I don't look for new information about my breakfast habit. Habitual behaviors also require little effort and are often performed in parallel with other behaviors. (I generally listen to a podcast while eating, for instance.)

Because habits are triggered by situations or circumstances, such as walking into the kitchen at a certain time of day, they are devilishly hard to change. You may be motivated and intend to change a bad eating habit, such as not ordering French fries and a hamburger when you go out for lunch with friends. But you repeat the behavior nonetheless, often without even thinking. In fact, whatever consciously reasoned intentions and desperate desires to change you may harbor are likely to be defeated from the start. Or, as psychologists put it, our surrounding's "automatic activation" of relatively mindless behaviors enables habits to persist despite our best intentions.[25]

What *does* change a habit? Someone trying to replace climate-unfriendly with climate-friendly eating habits has several options. They can change the situation that triggers the bad eating habit. Or, if unable to change the triggering situation, they can consciously plan for how they will respond to that situation, including deliberately practicing thoughts that will avert the unhealthy behavior. And once they've tried a new behavior, they can repeat it over and over until it eventually becomes a new habit.[26] Habits are easier to break during a change in other daily routines, such as when people move to a new home, start a new job, or send their kids off to college.[27]

Let's dive deeper into this idea of trying to change the situation that cues up the habit, or situational cues. Situational cues include aspects of the context in which the habit occurs—the time of day, place, and who you are with. One way to change habits is to change some aspect of this context. When you drive along the highway, you stop at McDonald's, and . . . but wait. Perhaps you can stop at a different restaurant to break the burger-and-fries habit—a restaurant that offers fresh, tasty salads. Or maybe you can buy fruits and nuts beforehand and eat at a rest-stop picnic table instead of at McDonald's. The situation that cued up the bad behavior has been disrupted, and the new situation provides an opportunity to try out new behaviors.

Nudges are situational cues that have been intentionally designed to change behaviors. They are aspects of the environment—what we hear, read, and see—that alter people's behavior without using force or law.[28] Nudges are particularly suited to addressing food habits where people are not thinking deeply about their choices. A nudge could be a norm message—for example, a message to a friend about the growing number of mutual friends who are participating in "Meatless Mondays." Without a lot of thought, the friend decides to forgo eating meat on Mondays because that's what everyone else is doing.[29] Another type of nudge is changing the "choice architecture," or what food choices are readily available.[30] A school cafeteria can do this by putting the meat-free foods in front, where students are more likely to see and choose them. As my daughter was planning her wedding, she was thinking about whether to serve meat. Because she was concerned about pushback from family if meat was cut out altogether, she decided to place vegan options up front as guests went through the buffet line, with chicken (a relatively climate-friendly meat compared to beef) in a less convenient location. By planning the event so that the low-carbon foods were prominent, she hoped to nudge people toward the more climate-friendly options.

But when negative cues persist, how might we change our response? Imagine I enter McDonald's because I am traveling with a friend, and he wants to eat there. I know this in advance and have a long time to think about it as I am getting bored driving through Ohio. I consciously rehearse the scenario that triggers the habit—going into McDonald's—and imagine a different scenario: when I get to the counter, I will order a McDonald's salad. In my head, I use an "if/then" scenario: if I go into that restaurant, I will order a salad regardless of what my friend is ordering. As I am driving through Ohio, I might consider a second "self-regulation" technique that involves identifying a personal goal, in this case eating more vegetables. I then imagine positive outcomes of reaching that goal—for instance, feeling happy that I have improved my health and made a contribution to solving the climate crisis. I also plan a way to respond to the obstacles I will encounter along the way: if my friend asks why I'm not ordering fries and a burger, I'll respond by making a joke about becoming a righteous and virtuous vegan.[31]

You might be able to change your own habits, but can you change your friends' habits? Again, you might consider changing situational cues by replacing the restaurant where you normally dine together with a place offering more healthy choices or by serving more climate-friendly options when you invite friends over for dinner or brunch. For friends who are concerned about climate, you also can set up a way for them to report out any climate-friendly choices they are making (e.g., posting photos of vegan meals on Instagram) and then recognize them for their accomplishments. Finally, remember that new habits are only formed if the behavior is pleasurable and repeated over time.[32] Alas, that one sustainable tapas brunch won't do the trick. Instead organize plant-rich meals on a regular basis, make the meals tasty and fun, and applaud your friends and family when they try something new.

Access

Although many richer white folks in the United States think of veganism and vegetarianism as being limited to people who look like them and who flock to Whole Foods (which some refer to derogatorily as "Whole Pay Check"), we see increasing evidence that people from all ethnicities and walks of life are trying out new ways to retain their food cultures without meat. From Grass VBQ Joint in Stone Mountain, Georgia, whose chef was recently named *Southern Living* magazine's cook of the year, to the Texas vegan food truck called Houston Sauce Pit, African American chefs are converting a southern tradition into something healthier for their customers and the planet.[33] And vegan Puerto Rican and Colombian restaurants are sprouting up in New York and Miami. In Los Angeles, Vegatinos offers jackfruit tacos that taste like *al pastor* (spiced pork-on-a-spit from central Mexico), and vegan caterer La Venganza won LA's Taco Madness competition for best taco in Southern California. Reflecting the ethnic diversity of vegan eating, National Public Radio reported, "pop-up festivals organized around Vegan-Mex vendors in Southern California have become a local sensation. They usually occur in working-class Latino suburbs like Santa Ana, Ontario, Highland Park, and Whittier. Full-time vegans—or—white ones, for that matter—are a minority at these events."[34]

A growing rural-urban economic and social divide, with rural residents having less access to healthy, fresh foods, may explain why smaller percentages of rural people compared to city dwellers have recently lowered their meat consumption.[35] Of course, food deserts also occur in low-income urban neighborhoods. In promoting a plant-rich diet, being aware of whether people have access to healthy foods is important. When access is limited, working with local grocers and government to make healthy foods more available—often through supporting a food-donation or other nonprofit or government program like Farm to School—is critical.

Network Climate Action: The Strength of Strong Ties

Fully 52% of people who are at least mostly vegan or vegetarian say that some or most of their closest family and friends also follow vegan or vegetarian diets. Just 8% of people who are not themselves vegan or vegetarian say the same.

—Cary Funk and Brian Kennedy, Pew Research Center scientists, *The New Food Fights*

During her Cornell climate online fellowship in spring 2019, Diana Crandall was trekking across the southern and western United States with her fiancé in their van. She was twenty-six at the time and chose as her climate action a plant-rich diet. She chose other people living full time in recreational vehicles (RVs) as the network to eat more plant-rich meals alongside her. Her goal was to influence a new generation of younger "vanners" through social media, but she also wanted to reach traditional RVers by publishing an article on plant-rich diet in an RV magazine. Crandall was experienced with social media; she had worked on the social media strategy for the massive March for Science demonstrations in 2017. She also felt she understood vanners, being one herself. To boost her network climate action (eating less meat), she offered a "reward" for anyone posting a picture of a vegetarian meal on Instagram—a fifty-dollar gift certificate for purchases at Kampgrounds of America (KOA), a common overnight stop for RVs.

Crandall launched her social media campaign. . . .

No takers.

So, she expanded her network to include her Instagram followers and upped the reward to $100. Still no takers.

Crandall encountered even more challenges beyond mere disinterest. One RVer belittled her looks on Facebook. Others questioned the reward. Why would someone they didn't know offer them money for making and posting a meal with a bunch of vegetables? Perhaps she was running into the problem discussed in an article in *The Guardian* titled "Why Do People Hate Vegans?"—that is, our annoyance with people who appear to "virtue-signal," or advertise their virtue with the seeming intent of forcing their moral behaviors on others.[36]

On one hand, Crandall knew that external rewards (like a KOA gift certificate) can influence behaviors. On the other, many other factors influence what we eat, and those factors seemed to be foiling her valiant efforts. Food choices convey cultural values and in some cases commitment to faith. They are more likely to be influenced by family members than by friends,[37] and they may be influenced by housemates or others with whom we frequently cook and eat. They are influenced by what we believe is healthy and what meals we know how to cook. And they are influenced by what's available for purchase nearby—for instance, at KOA campgrounds. What's more, people don't always make food choices consciously, instead simply eating what we have a habit of eating or what tastes good.[38] How much power does anyone have to counter all these forces on others?

When I first envisioned the Cornell Climate Online Fellowship, I imagined we would all become influencers. We would simply post scrumptious-looking meals created with foods showing a palette of colors that would appeal to any palate, and the habit of eating plant-rich meals would go viral among our Facebook and Instagram followers. But that didn't happen.

After listening to Diana Crandall's story and feeling her frustration (I myself was trying to influence my office colleagues to pay for air travel offsets and was not seeing that behavior go viral either), I came across a book by the sociologist Damon Centola called *How Behaviors Spread: The Science of Complex Contagions*.[39] Centola starts with the example of contagious diseases. In 2014 a single individual with measles visited Disneyland and infected over one hundred other people. The infected people didn't know each other—they likely didn't even touch each other. Yet the measles virus spread rapidly to those who just happened to be nearby. In social network terms, the Disneyland visitors were a random network with extremely weak ties—yet they still got sick. Today we see false ideas such as stolen elections also spread through weak ties, especially when they are promoted by celebrities such as Donald Trump. The sociologist Mark Granovetter captured this phenomenon in his classic 1974 paper titled "The Strength of Weak Ties." Simple ideas spread in weak-tie networks, where we encounter people with different views and resources. Our eyes are opened to diverse perspectives and opportunities that we wouldn't normally confront through our strong-tie networks of family and close friends, who think and act more like us.[40]

But eating a plant-rich diet is neither a contagious disease nor a simple idea. It is a complex behavior determined by a complex set of factors. This is why we need to understand the science of complex contagions.

According to Centola, complex behaviors *can* spread through networks—but not through weak-tie networks like Crandall's RVers.[41] Instead, we might expect another Cornell climate fellow's approach to be more successful in influencing others to convert to a plant-rich diet. Fatima Delgado was a Cornell climate fellow from Ecuador. During the fellowship she was completing her PhD at a university in Barcelona, Spain. Like Crandall, she chose plant-rich diet as her network climate action and invited a group of twelve close friends over to her apartment for "sustainable tapas" brunches. Being friends, they had strong ties compared to the RVers in Crandall's online network. Because they knew each other, they might

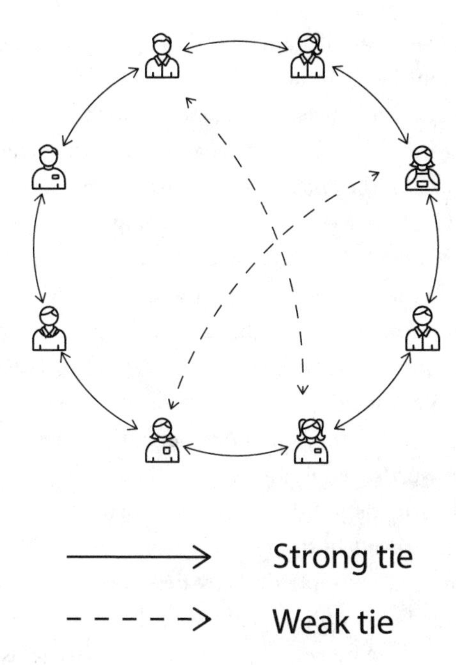

Strong and weak ties in a network. Diagram by Xiaoyi Zhu.

even reinforce the message about sustainable eating among friends, repeating it during and after the brunches. This meant that each member of the network might hear the plant-rich diet message not only from Delgado but also from other friends. And because Delgado had invited them over for sustainable tapas several times, they would hear the plant-rich diet message from Delgado and their other friends repeatedly. Unlike Crandall's weak-tie network, Delgado's was a network whose members knew each other well; they shared strong ties.

Having read Centola's work about how behaviors spread through strong-tie networks, I coined the term "network climate action" to capture the notion that in changing our own behaviors—that is, in taking a climate action such as reducing how much meat we consume—we can influence others to whom we are close to take that action alongside us. As Delgado's story illustrates, network climate action entails three steps. First, identify the most effective actions you can take to draw down greenhouse gas emissions. Second, take those actions yourself. Third, think of creative and fun ways to draw in close family and friends—that is your tight social network—to take those same actions alongside you.

Imagine a network where you have strong ties with a few close friends. Each friend also has strong ties not only with you but also with several other friends. And a sizable portion of your friends' friends have ties with you and with each other. In short, you are part of a *clustered* network. A person in a clustered network has redundant or overlapping ties—your friends are likely to also be the friends of others in the network. And if someone takes up a new behavior—let's say, eating vegan—you might hear about it multiple times from multiple friends. The news of your friend becoming vegan circulates around a group of close friends.

We can see how *information* about veganism would spread in that network, but what would it take for the *practice* of veganism to spread among those friends? Likely, if one friend is already vegan, she is not going to change anyone's behavior. But what if several friends are vegan? Not only would the members of that friend network hear about veganism more often, but they would also be more likely to be invited to vegan dinner parties within their group of friends. Based on computer models and experiments with real-life social networks, Centola has shown that complex behaviors, such as becoming a vegan, spread faster through these clustered networks than through random networks where people don't know each other. In other words, people are more likely to adopt new behaviors if their friends and friends' friends (who also may be their friends) are adopting that behavior. This is the *strength of strong ties.*

You might think that a couple of weak ties to a distant person, whom you or your close friends know only vaguely, would speed up the spread of vegan eating across a network. Why wouldn't it help spread veganism, even just a little, if a person had a weak tie with an individual on the other side of town? After all, that distant individual can spark behavior change among their close friends. But here's the catch. If you share your vegan eating with someone not close to you, that person may be inclined to push back with their own opinions. Whereas a close friend might feel free to talk with you about any concerns they have about your eating habits or even your evangelism for vegan eating, someone

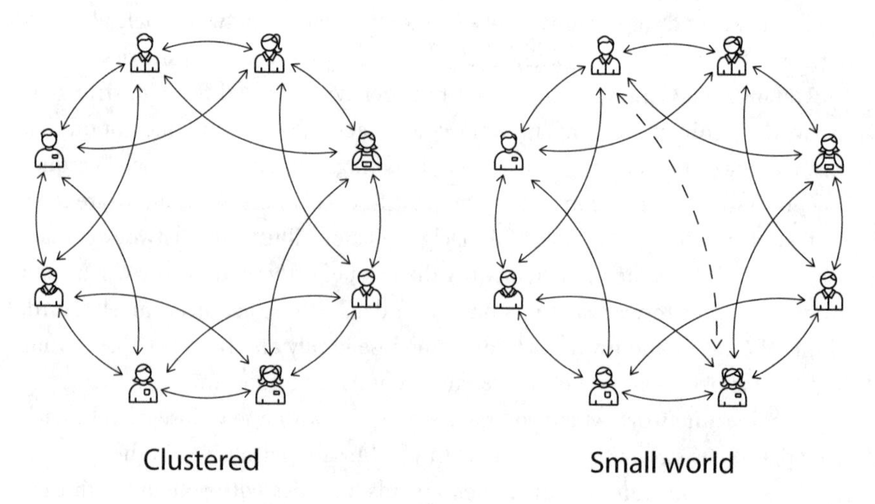

Clustered and small-world networks. Diagrams by Xiaoyi Zhu.

with whom you are only weakly connected may feel entitled to belittle you. The weakly connected person's derision may then impact close friends who were just on the verge of starting to go vegan—and turn them away from adopting this new eating behavior to avoid being ridiculed or sanctioned. Because of this push-back from individuals only loosely connected socially, the spread of complex behavior through networks with both weak and strong ties—or so called "small-world networks"—can actually be slower than the spread of complex behaviors through clustered networks with only strong ties.[42]

What else might account for the weakness of the weak ties in Crandall's network? Let's assume that she shares the RV identity with other RVers—being

Identity

Different kinds of identity can exert different influence.

Personal identity: How an individual thinks of themselves, especially to distinguish themselves from others (e.g., mom, church-goer, composter, plalker)[43]

Social identity: Aspects of our self-image that are shared with a larger group (e.g., environmentalist, millennial, vegan)[44]

Collective identity: A group or social identity focused on members' associations with, and contributions to, social movements (e.g., climate activist)[45]

How invoking identities can increase persuasiveness: We all have multiple identities—for example, a woman, a vegan, an Asian American, or an environmentalist—some of which are more salient at a given time. One way to influence behavior is to create a situation where a particular identity becomes more salient or significant.[46] We might share photos of a climate protest to make a climate activist collective identity more salient, which in turn might influence people to advocate for climate causes. Researchers have found that when people describe their eating and other consumption choices, they often refer to their identities as parents, children, siblings, friends, workers, retirees, churchgoers, or activists, any of which could be invoked to try to influence eating behaviors.[47] For example, a person who strongly identifies as a mom could be invited to join a mom's healthy eating group that shares plant-rich recipes and meals with their kids.

an RVer is central to how they describe themselves. But many of the RVers and Crandall don't share other salient aspects of their identity, such as gender, political party, and exercise habits. People are more likely to adopt healthy behaviors if they are the same gender and have similar levels of obesity. Crandall is female, young, liberal, thin, and physically active. Perhaps the other RVers were mostly male, older, political conservatives, and a bit out of shape after driving their RVs so many miles. This lack of shared salient characteristics could have thwarted her attempts to convince RVers to adopt a plant-rich diet. Not only network configuration—like weak and strong ties—but also the identities of people within a network influence the spread of complex behaviors.[48]

One reason behaviors spread more readily through close networks is that they provide opportunities for "deep engagement." Instead of a simple appeal, a person is asked to connect personally with the climate issue. This might entail a discussion around the dinner table about how caring for the climate aligns with a person's religious faith, ethnicity, or desire to leave a legacy for their children and grandchildren.[49] It's true that sometimes people change their attitudes or buy something without much thought after watching catchy advertisements or simply because they like a celebrity messenger. But a person who thinks deeply about an issue is more likely to change complex behaviors or attitudes for the long term.[50]

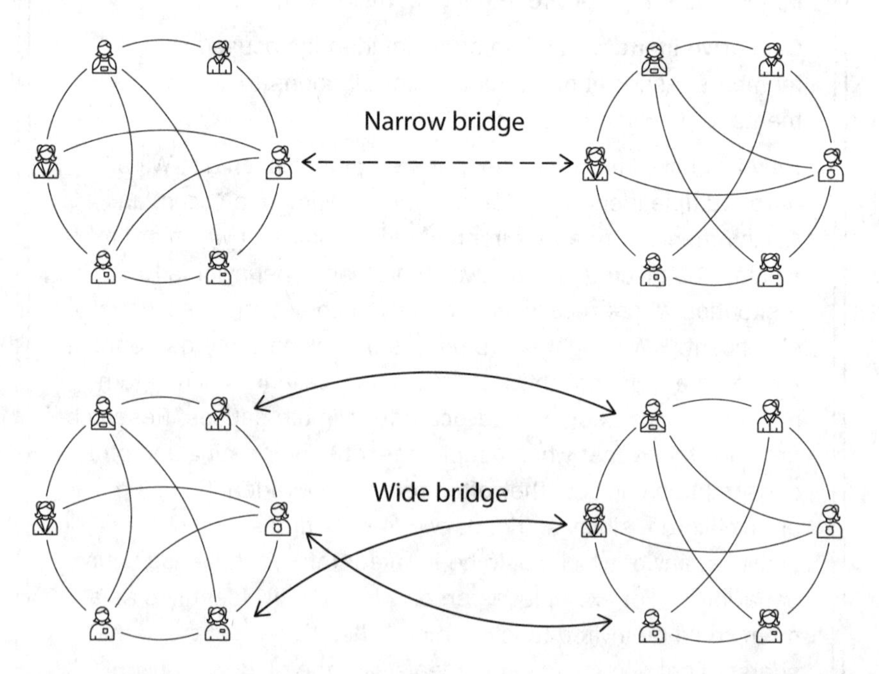

Narrow bridge with one person spanning between networks. Wide bridge with several people spanning between networks. Diagrams by Xiaoyi Zhu, redrawn from Centola, *How Behavior Spreads*, 2018.

Once a behavior has spread through a tight or clustered network, how might it spread from that network to another? Here again strong ties come to bear. Imagine two sororities on a college campus. A "bridge," or woman who has ties with other women in both sororities, can spread information, ideas, or even a head cold. Single individuals with ties in two sororities or networks serve as "narrow bridges" between the two groups. But let's say one sorority starts serving plant-based meals. Unless there are "wide bridges"—or several individuals who span both sororities—it's unlikely this complex behavior, plant-rich eating, will spread to the second sorority. Women in the second sorority need to encounter plant-rich meals from multiple women who bridge both sororities. In other words, for a behavior to spread from a so-called seed neighborhood to another neighborhood (or from a network or sorority to another), multiple individuals in the first

Under the Influence

Too often we commit the "fundamental attribution error." If we have friends who smoke, eat a lot of snack foods, or don't exercise, we might blame our friends' unhealthy behaviors on weakness of character. But we should beware of attributing others' behaviors to their character or personality without accounting for external influences, including the social influence of friends, family, and colleagues. Policies that prohibit smoking may be less important because they stop single individuals from lighting up or protect their friends from secondhand smoke. Rather, their importance may lie in the fact that they expose individuals to more people who have quit smoking and thereby to their influence.[51]

When you put solar panels on your roof, bike to work, or buy an electric car, you may also influence neighbors and friends who see one more person or family like themselves adopting a climate behavior and who want to follow the trending norms in their community.[52] In fact, copycat behavior is so prevalent that the economist Robert Frank has written an entire book, *Under the Influence: Putting Peer Pressure to Work*, arguing that the most important influence of a person adopting a climate behavior is not that their individual action makes a dent in greenhouse gases in the atmosphere. Rather, it is that they put their peers "under their influence" and thus spread their behavior to others.

neighborhood need to have ties with multiple individuals in the second neighborhood. These multiple ties become wide bridges between the two networks. In this way, women in the second sorority hear about the new behavior multiple times and from multiple messengers with whom they share a common identity as a sorority member.

Characteristics of Behaviors That Are Likely to Spread

Even when multiple messengers are sharing multiple messages about a new behavior, some complex behaviors spread, whereas others stagnate with just a few people adopting them. What are the characteristics of behaviors that spread? Centola has discerned four factors that make it more likely behaviors will spread through a network: strategic complementarity, legitimacy, credibility, and emotional engagement.[53]

Recently, I have begun responding to "action alerts" from my local chapter of the nonprofit Climate Reality Project. On weekends I leave messages for state and federal representatives urging them to support climate legislation. As I sit alone in my living room dialing their numbers, I wonder if my calls amount to anything. But then, on a recent Sunday, I participated in my first tweetstorm. Climate-concerned folks across the United States were asked to like and retweet tweets with the hashtag #ActOnClimate. At the same time as I was racing down the Twitter feed to retweet as many times as possible during our one-hour tweetstorm, I was on a Zoom call with about fifty other climate "retweeters." We were chatting and checking to see if #ActOnClimate was trending in the top thirty hashtags. Perhaps unsurprisingly, even though our posts and retweets were seen by nearly seven and a half million people, we were beat out on Twitter by *Sunday Night Football.* But seeing so many other people taking action alongside me felt much more worthwhile and fun than calling alone on the weekend. The fact that so many people were participating in the tweetstorm made our behavior more valuable; we complemented each other's actions. Centola calls this phenomenon—of a behavior becoming more valuable as more people adopt it—*strategic complementarity.*

When I lean over to pick up trash on my way to work, I am half hoping no one is watching because I don't want passersby to think I'm some crazy old radical hippie environmentalist. But I also think that seeing me may make the behavior appear more legitimate and make others a bit more likely to take up the plalking habit themselves. The other day, I was a little surprised to discover that Jonathan Zacharias, a student in my Climate Solutions class who is a six-foot-four baseball player, is a vegan—a behavior that, like plalking, is out of the mainstream. I also discovered that another student, Michelle Zheng, used WeChat's VegPlanet platform to help her parents, who live in the United States but speak mostly Chinese, see that veganism was gaining popularity in China. As more and more people adopt these behaviors, the less we'll fear being ostracized for engaging in them. The habits may eventually even become the norm. I have seen this transition at professional environmental educators' conferences where I used to have to ask for vegan dishes, but then they became the default choice. Vegan eating has obtained *legitimacy*, the second factor influencing behavioral spread in networks.

When the topic of athletes and vegan diets came up, my baseball player student suggested that I watch *Game Changers*, which is touted as "a revolutionary new film about meat, protein and strength" featuring Arnold Schwarzenegger, Jackie Chan, Lewis Hamilton, Novak Djokovic, and Chris Paul.[54] If trusted top athletes and adventure movie idols are vegan, then student athletes may conclude that practicing veganism won't impair their performance after all. I follow a similar logic when I participate in advocacy actions suggested by the Climate Reality Project. The organization was started by Al Gore, and lots of people I trust are members, which lends *credibility* to a particular action. I trust them so much that I don't feel I need to read the background information about each policy they tell me to support, which saves a lot of time, making it easier for me to participate in advocacy actions. We mimic people we trust and who have credible information that we lack.

Another Cornell student in our Climate Solutions course would buy his friends the first round of drinks at each Friday's end-of-week festivities—but only if the friends ordered vegan snacks first. Assuming everyone was having a good time, eating vegan while consuming beer could become "emotionally contagious." Climate fellow Chandra Degia, who is a professor of conservation behaviors and the host of a Jamaican music radio show, also realized the importance of *emotional contagion* when she corralled musician friends to produce a fun reggae music video celebrating "One Day a Week, Eat Less Meat."[55] We are more likely to adopt a new behavior if we feel positive emotion or excitement.

How does all this fit together? Behaviors spread through clustered networks where we hear about the new behavior multiple times from multiple messengers. As the behavior becomes more common, barriers such as limited vegan choices on restaurant menus decrease and the behavior becomes easier to enact. As more people adopt the behavior, it gains legitimacy and credibility. This is especially true if people trust each other and share salient aspects of their identity, such as being a health- or planet-conscious person, an athlete, or a reggae fan. And as people become excited about the new behavior, it catches on even more.[56] Eventually the behavior has crossed a threshold or tipping point, where it becomes the accepted behavior or societal norm.[57]

Norms and Influence: Everyone's Doing It!

> **The surprising power of social networks is not just the effect others have on us. It is also the effect we have on others.**
>
> —Nicholas Christakis and James H. Fowler, social scientists, *Connected*

Norms—or agreed-upon standards of behavior—are an important part of the puzzle explaining social influence. Norms of eating behaviors can be communicated by health magazines in the waiting room at the doctor's office, by TV celebrities such as Martha Stewart and Oprah Winfrey, by online fitness influencers such as Simeon Panda or vegan TikTok megastar Tabitha Brown,[58] or even by politicians deriding climate activists. "They want to take your pickup truck! They want to rebuild your home! They want to take away your hamburgers! This is what Stalin dreamt about but never achieved!" ranted Donald Trump acolyte Sebastian Gorka, referencing American norms while deriding politicians who suggest the need to change our consumption practices to protect the climate.[59]

Norms can also be communicated by other people's behaviors, including how they eat.[60] If I am eating with people I admire who happen to be vegans, I may follow a vegan social norm because I want to be approved of or liked by them, or I may even want to belong to their social group. Social norms influence eating because they are associated with how people judge each other.[61] We might mimic what friends or associates say and do, not just because we value their expertise but also because we want to be part of a group whose members like and approve of us, and that group even may be part of our identity.[62]

And it works both ways. Just as norms influence behaviors, behaviors influence norms. Over a period of thirty-two years, scientists tracked the risk of becoming obese if your friend is obese. It turns out that if a person identified an obese person as their friend, their chances of becoming obese increased by 57 percent.[63] And if the friendship was perceived as mutual (not only did the individual identify the obese person as a friend, but the obese person also identified the individual as their friend), the individual's risk of obesity increased by 171 percent. What's more, studies show our behaviors can spread to our friends' friends and even our friends' friends' friends, even if we don't know them. The researchers attribute these results not literally to the spread of obesity per se

"Maybe the rabbits are right. Maybe carrots are delicious."

but to the spread of norms and behaviors that impact our weight (e.g., norms about exercise or diet) as well as to the esteem people bestow on their friends.[64] If I think highly of a friend and see that she doesn't get exercise or have a healthy diet, I may conclude that not exercising and eating junk food is okay—that these behaviors are in fact the norm among my friends. I may be less likely to go for a jog or resist those McDonald's fries.

Studies show that knowing what people *do*, or descriptive norm messages, are usually more effective at influencing behaviors than telling people what they *should* do, or injunctive norms. So, if we know that 70 percent of people in the United States are vegan, we are more likely to reduce meat and dairy consumption than if we are told we should be vegan. The problem arises when the behavior we are trying to encourage—like veganism—is not common. In reality, only about 3 percent of Americans are vegans, and about 5 to 6 percent are vegetarian.[65]

This is where dynamic norms come in. A Web post titled "Exploring the Explosion of Veganism in the United States" provides an example of a dynamic norm message. It reads, "Veganism in the US has grown from obscurity to become a mainstream part of the American diet. Since 2004, the number of Americans turning plant-based has reached 9.7 million people, growing from around 290,000 over a period of 15 years."[66]

Social Norms

Social norms are commonly accepted rules about how people in a society should behave. Three types of norms have been used to try to influence behaviors:

Descriptive norms tell what people *are doing*.

Injunctive norms tell what people *should* do.

Dynamic norms (also called trending norms) tell what *more and more people are doing*, or what behaviors are becoming more common.[67]

A study at Stanford University compared the effects of descriptive and dynamic norms on students and staff at a university cafeteria. The heart of the descriptive norm message read, "30% of Americans make an effort to limit their meat consumption. That means that 3 in 10 people eat less meat." The dynamic norm message read, "In recent years, 3 in 10 people have changed their behavior and begun to eat less meat." The researchers found that the dynamic norm message doubled the percent of students and staff who ordered meatless lunches compared to the descriptive norm.[68]

Opinion Leaders, Influencers, and Affluencers

Could the fact that Diana Crandall was not an opinion leader in her RVer network have worked against her? Opinion leaders have connections to many others in a network, can influence followers' attitudes, and are often looked to for advice. Opinion leaders may or may not be innovators themselves; in fact, they generally reflect the norms of their group. But if opinion leaders are convinced to adopt new ideas, attitudes, or behaviors, they have a powerful ability to spread these new ideas, attitudes, and behaviors among those in their social network.[69]

Key to the original concept of opinion leaders was that they had strong interpersonal ties with those whose opinions and behaviors they influenced. As we have moved to more and more online and social media communications, we have come to trust certain politicians, journalists, and other elites, and we take our cues from them, including our opinions about climate change.[70] Celebrities, such as actors, musicians, and athletes, may also influence what issues we pay attention to and in some instances our political opinions.[71] The line between celebrities and social media influencers is blurry, but influencers tend to be known for the products, opinions, or lifestyles they sell, whereas celebrities are known for their music, acting, or athletic performances or other body of work.

Influencers persuade people to buy products and thus have figured out how to change people's behaviors. They do so by leveraging social media to create a reciprocal relationship with followers. For example, they like and share followers' posts, ask followers for suggestions, and cocreate YouTube or Instagram content with followers. By focusing on a particular lifestyle and audience, engaging as a "normal" person in a running dialogue with their followers, and sharing their own seemingly authentic messages, influencers also build credibility and trust. Although we may never meet them in person, influencers create what are called "trans-parasocial relations" with their followers. They are perceived as friends, which enables their purchasing behaviors to spread among their followers.[72]

The Kardashian family are both celebrities and influencers. They invited the world into their daily lives for fourteen years through their weekly TV series *Keeping Up with the Kardashians*. Their ability to form intimate connections with their audience, along with their celebrity status, beauty, and wealth, may help explain their social influence. When Khloé Kardashian modeled her new Doll Beauty Lipstick on Instagram, it was sold out within four hours, and once stocks were replenished, it sold out again the next evening. The lipstick brand's website then crashed after being inundated by a 520 percent increase in traffic.[73] At a minimum, influencers can influence lipstick-buying behavior.

Whether Khloé Kardashian or another celebrity or influencer could persuade followers to adopt a more complex behavior, such as switching to a plant-rich diet, is doubtful. But a celebrity might be able to put the notion into someone's head. That person might then talk to friends and family about plant-rich meals, perhaps eventually leading the individual and their group of friends or relatives to try healthy recipes. Those friends and family members might then be able to spread the behavior where Kardashian has no influence—that is, among people who are not her followers. And when a person is considering a new behavior suggested by a friend, they might look to influencers and celebrities to reinforce the legitimacy of that behavior.

Kim Kardashian and Kanye West, who purportedly hired private firefighters to protect their sixty-million-dollar mansion during the 2019 climate-change-induced wildfires in California,[74] are not just celebrities and influencers but also "affluencers"—that is, affluent elites who are keen on trying and recommending new products and services. The affluent tend to be influential out of proportion to their size as a demographic class: with money to spend and the ability to experiment, they're often the earliest adopters of products and services.[75] Their luxury consumption has an oversized carbon footprint, which is multiplied if they influence others to take up their profligate lifestyle; witness the rush to copy Jeff Bezos's private launch into space.[76] In societies such as the United States, where the rich often display conspicuous consumption, others copy them by buying ever more expensive yachts, watches, and wines.[77] Here we need to turn to legislation that eliminates vast wealth inequities in order to promote a more just and climate-friendly society. One way to help make this happen would be to support an organization fighting for social and economic equity, such as the People's Justice Council.

Action among Friends

Regardless of how persuasive celebrities and influencers can be, network climate action requires strong ties to be successful. For one thing, many of the important climate solutions are complex behaviors we often do in our homes, where "seeing the change you want to be" (and "being the change you want to see") happens among small networks of friends and family. Perhaps more important, most of us will never have hundreds of thousands or millions of followers. But we can influence people in our close friend, family, and coworker networks by sharing our climate-friendly meals—and more broadly our climate actions—repeatedly over time. I'll close this chapter with a couple examples from our Cornell climate fellows.

Max Schubert, a German undergraduate student studying forestry at the time of the Cornell fellowship, invited members of his professional network to cook meals with him. He was hosting a meeting in Bonn for seventeen leaders from the International Forestry Students' Association. He decided to go all out—everyone was going to shop, cook, and eat only vegetarian for the two weeks they were together. (They also only took public transport and paid for air travel offsets.)

Because Schubert's colleagues were in a new setting away from their daily routines, he did not have to worry too much about eating habits and traditions. Instead, by engaging his network in cooking vegetarian meals, Schubert was addressing another barrier to embracing a plant-rich diet—that is, knowing how to cook vegetarian.[78] If your friends are interested in reducing their meat consumption, then recipes and modeling how to cook can give them the knowledge—and the feelings of agency or self-efficacy—that they need to get started.[79] Knowing how to cook vegan meals is a form of action-related knowledge—that is, knowledge about how to do something. We often stop after sharing systems knowledge, such as how meat consumption impacts climate or heart health, but that alone is unlikely to change anyone's behaviors.[80] Showing people *how* to do something is generally more impactful in changing behaviors than telling them why the behavior is important, especially if they already have an interest in changing that behavior.

Because Schubert met with the other students for only two weeks, they may not have experienced vegetarianism long enough to change their meat-eating habits. Also, the students returned to their home countries and to their existing networks of family and friends, with their own eating traditions and identities and where vegetarianism was probably not the norm. Once people return to the old situational cues—including the housemates they cook with and the places where they eat—they may revert to old habits. Thus, showing people how to engage in climate-friendly behaviors—rather than assuming that they already know—is a good first step, but forming a new eating habit may require more than an intense couple weeks of preparing meals together. For Schubert's colleagues

PANIC Principles of Social Mobilization

Harvard University policy professor Todd Rogers and colleagues compiled five principles that can be used to influence behaviors. These principles can be remembered using the acronym PANIC, which is derived from the first letter of each principle. The principles capture much of what we've discussed in this chapter, including social connections, norms, and identity. I am not sure if it's because of the catchy acronym or simply because the principles work so well, but my students have found the PANIC principles and acronym particularly helpful in their network climate actions.

***P**ersonal:* Initiate more personal interactions between people who can relate to one another.

***A**ccountable:* Ensure that reputation-relevant behavior is observable to others.

***N**ormative:* Convey what relevant people think others should do as well as what relevant people actually do.

***I**dentity relevant:* Align behaviors with the ways people actually see themselves or would like to see themselves.

***C**onnected*: Leverage people's networks of relationships and the platforms that maintain those networks.[81]

to change their eating habits over the long term, they would need to think about ways to change the situational cues that trigger meat-eating behaviors once they returned home. They would also need to think about where they can get support from other vegetarians who have dealt with incorporating meat-free meals into the culture and eating traditions of their family and community. And Schubert might want to maintain his forestry vegetarian network online, reinforcing their identity and plant-rich cooking through ongoing communications and recipe sharing, especially if he can make it fun.

Let's close this chapter with the story of Avinash Acharya, another climate fellow, who works for an international nonprofit that collaborates with businesses on climate change and global sustainability. Acharya and many of his friends and colleagues are members of India's educated middle class. They are committed to sustainability and seek out sustainability actions they can take. During our climate fellowship, Acharya shared the Project Drawdown website with twenty-five Indian urban professionals in his network. Knowing that his peers would want to have a say in their climate actions, he asked them to choose their own Project Drawdown actions—as many actions as they could reasonably manage. He kept a running spreadsheet of their actions, along with the amount of carbon emissions they were avoiding. By week three of Acharya's climate campaign, the combined efforts of his friends and colleagues were approaching five tons of CO_2 equivalents of emissions avoided.

A member of Acharya's climate action network shared to their "Climate Save November" Facebook group:

> Yes, we are a privileged middle class, but it is time we recognize and use our privilege for lifestyle changes because we can do it—and doing this does not make us judge people who, for cultural economic reasons, are unable to. . . . Yes, big corporations must be held accountable for the havoc they wreck [*sic*] on the environment, but let us continue questioning them all the while doing our little bit where we can. And yes, we believe that even if it is only 25 people in a city right now tracking their daily emissions, it is a start. #climatesavenovember should be "climatesaveallyearround."

What was the key difference between Crandall's RVer network and Acharya's friend-and-colleague network? For one, Acharya's network had stronger ties. Members also shared similar backgrounds, environmental values, and professional identities, and they likely trusted each other more than did members of the online RVer network. Because they were friends and colleagues, they could reinforce each other's behaviors on an ongoing basis.

In choosing your climate action network, consider choosing family members and friends who are already primed to explore climate solutions and thus easily nudged. Such "nudgeable" networks, like Acharya's educated, urban Indian millennials, adopt climate behaviors when provided with knowledge about how to reduce their climate footprint and asked to join a climate challenge. Many in the United States have spent years wringing their hands about climate deniers—people who are, themselves, facing extinction as the climate crisis rears its ugly flood-and-fire-spewing head and extinguishes more and more lives. It's important to remember that the climate justice movement does not require every last contrarian to come along at this stage of voluntary action. Instead, the effort should be to enlist people who are already interested and waiting for the right nudge to push them into initial action and more impactful action. As public opinion, behavioral norms, and eventually policies change, more and more people will jump aboard the climate-solutions train.

• • •

EAT: Don'ts and Dos

1. Don't think that you have to become a vegan to make a difference. Going vegan would be great, but simply reducing meat and dairy where it works for you and your family is still helping reduce greenhouse gas emissions.
2. Don't go it alone. Your most important influence will be sharing your plant-rich meals with family and friends, helping them to understand why you have chosen to reduce meat and dairy, and showing them how to prepare tasty plant-rich dishes.
3. Don't waste your energy on stubborn cases. Not everyone is open to changing their eating habits. Focus on family and friends who will be most open to plant-rich eating.
4. Do be aware of barriers to plant-rich eating, including friends and family thinking it costs more, is time-consuming, can't provide enough nutrients, or doesn't taste good. Demonstrate how plant-rich meals can be cheap, relatively easy to prepare, healthy, and tasty.[82]

2

GLEAN

A year after completing our online course, retired professor Mark Schlesinger was still pursuing his network climate action. He wrote to a friend trying to persuade her to compost:

> Fran and I were composting for years before we moved here. The reasons were manifold. It was convenient, it reduced waste in the house and garage, saved money on fertilizer for our garden, provided rich footing for our vegetables and flowers, and—and this was most important to me—reduced carbon emissions, even if it was a proverbial drop in the bucket. . . .
>
> You asked me personally, and it is a personal priority. As a father and grandfather more than a little concerned about the world our children inhabit and our granddaughter will experience, I want to do whatever I can to make that world a healthier place, even if it's just a little bit, which our composting effort would be. I believe we let them down if we don't now act on their future behalf.
>
> I also believe that it can be done with a minimum of stress to people who are not interested and with maximum benefit to those who are.

For Schlesinger, composting is one small thing he and his wife Fran can do for their granddaughter, and it can be done easily and "with a minimum of stress." In contrast are more fraught food-waste actions that aim to goad the public and private sector into radical transformation. A paper titled "'Waving the Banana' at

Capitalism" portrays a scruffy New Yorker named David performing from inside a food dumpster as citizens and the media look on: "A bruised tomato acts as a vehicle to lambast labor practices on corporate farms; a discarded carton of eggs opens up an opportunity to attack factory farming; organic lettuce provides a medium for a tirade against 'green consumerism'. . . . The speech, which David calls 'Waving the banana' after one of his finds, closes with a call for a new social model founded in cooperation, communalism, and a 'gift economy.'"[1]

As a so-called freegan (an amalgam of "free" and "vegan"), David combines veganism with seeking out free food and other free goods and often uses performances such as dumpster diving to protest consumerism and capitalism.[2] He also aims through his performances to attract new adherents, who will join him in "waving the banana at capitalism" and demanding the transformation of our industrial food system.[3]

Curtailing food wastes encompasses a suite of practices, running the gamut from composting in your backyard to diving into a dumpster to recover food that's still edible, with the crucial common denominator being that less food is tossed out. But how might these and other food-waste practices go beyond the lone individual who decides to buy only the food they can eat, the families like Schlesinger's who dutifully tend their compost piles, or the individual radical anticapitalist who liberates food from their neighborhood grocery store dumpster? How can those individuals add up to a movement, a sum greater than its parts?

One way to answer this question is to consider the connections people make (or fail to make) between their food-waste and other practices, such as cooking and shopping. The answer also lies in whether people consider not only sustainable food-waste practices but also factor in sustainability and social justice when choosing what to eat and where to purchase food. And it depends on whether people like Schlesinger who start with composting in their backyard or like David who wave a banana fished out of a dumpster have the stick-to-itiveness, the dogged perseverance, to garner the support of neighbors and local elected officials.

Why Reduce Food Waste?

> **Throwing away food is like stealing from the table of those who are poor and hungry.**
>
> —Pope Francis, World Environment Day, 2013

After farmers' backbreaking labor to produce food, over one third of what farms yield never reaches our stomachs.[4] The amount of food people waste grew over the last century, likely due to food's relatively low prices and the fact that most of us have not experienced food shortages in our lifetimes. Another factor contributing to food waste is the changing roles of women, many of whom no longer have the time or skills to create tasty dishes from leftovers.[5] In the mid-2010s the United States wasted over $218 billion on growing, processing, transporting, and disposing food that was never eaten.[6] This amounts to about 35 percent of the total food produced and 2 percent of total US GDP.[7] If global food waste were a country, it would be the third largest greenhouse gas emitter after China and the United States.[8]

Food lost from the farm to the household is estimated to generate 8 to 10 percent of total human greenhouse gas emissions.[9] But wasting food is not just a contributor to the climate crisis. It's also an environmental, social, and economic problem. More land and water must be diverted to agriculture to make up for food losses. Food waste costs money for everyone from the farmer to the grocer, restaurant owner, and consumer. And good food is tossed out while many go hungry.[10] The amount of food wasted reflects global economic disparities and is a global justice issue. Food loss per capita is 6–11 kilograms per year in Central and West Asia and North Africa but over ten times that amount (95–115 kilograms per year) in North America and Europe.[11]

Yet wasting food is also a solvable problem. In the Nordic countries, nearly 90 percent of food waste is collected and transported to incinerators to generate energy or to anaerobic digesters to produce biogas. Some discarded food is repurposed for human and animal consumption or composted and returned to the farm to grow more food.[12]

But what about actions we can take so that food never reaches the curbside for pickup or rots on the way to market? It's way more sustainable for people to eat food than for food to be converted to energy or compost.

Far from Scandinavia lies a rural community in Zimbabwe where most households can't afford to waste food. Zimbabwe's food-waste problem is about getting farmers' hard-earned produce to the market before it spoils. Respect Musiyiwa, a Cornell climate fellow from Zimbabwe, is trying to find solutions. Musiyiwa's father passed away due to AIDS when he was ten years old. When his mother, who had struggled for years to pay his school fees, also succumbed to AIDS ten years later, Musiyiwa had to abandon his university studies. He returned home to his rural community in the Chiweshe District. There he committed himself to becoming a social innovator and entrepreneur to battle rural hunger and poverty. After his initial efforts to mobilize youth to fight against marginalization were stymied by political instability, in 2012 he launched a social enterprise, the Centre for Agro-entrepreneurship and Sustainable Livelihoods (CASL) Trust. While a Cornell climate fellow, Musiyiwa worked on a project to dry locally produced fruits, vegetables, and herbs using solar energy. Farmers would earn income from selling the preserved foods, which otherwise would have rotted due to lack of refrigeration and unreliable transportation on the way to urban markets. By 2018, working with over 150 small farmers and two farmer groups, CASL Trust had processed upwards of nine thousand kilograms of fruits and vegetables. As part of his network climate action, Musiyiwa convinced two community groups to replicate his team's solar drying project. Reflecting on his work, Musiyiwa wrote, "The people in my area are marginalized and poor and have little time and resources to face crises, hurricanes, and droughts. In order to be resilient and at the same time to help solve the root of the problem (climate change), reducing food waste is important."[13]

Unlike in poorer countries like Zimbabwe, where most food waste occurs before products reach the market, food in richer countries is most often wasted where it is consumed—in households.[14] Household food waste includes parts of food most people consider inedible, such as orange peels and nut shells—so-called unavoidable waste. More concerning is "avoidable waste," or edible food that is thrown out, such as the slimy lettuce forgotten in the back of the fridge, bread that turned green before we were able to get to it, and leftovers that our children refuse to eat.[15]

Food-Waste Hierarchy

Preventing food from being thrown out in the first place is the gold star of food-waste reduction.[16] Recognizing that, inevitably, an ounce of food produced is not necessarily an ounce of food consumed, researchers have called for a "circular economy" of food. In this closed-loop system, food that otherwise would go to waste, such as day-old bread, is donated to hunger relief programs or repurposed as animal feed, compost, or energy.[17] A simple hierarchy summarizes these food-waste actions: refuse to discard food, reuse food that's still good, and repurpose what's no longer edible.[18]

A number of creative grassroots initiatives have sprung up to avert the worst-case scenario of uneaten scraps being carted off to the landfill. Well-established food-rescue organizations such as City Harvest in New York City distribute unused but still edible items from grocery stores and restaurants to food banks and soup kitchens, thus reusing “waste” to feed the homeless and the poor.[19] A startup called Too Good To Go developed an app enabling consumers to buy “mystery bags” of whatever a restaurant has left over at the end of the day. Lest you think this is a marginal enterprise, Too Good To Go, launched in Denmark in 2015, grew from two hundred to eleven hundred restaurants in New York City from September 2020 to April 2021. And in January 2021 it raised $31 million for expansion to new cities and boasted over eighteen million users.[20] Although Too Good To Go has not yet reached my small city of Ithaca, New York, students in our Climate Solutions course are trying to bring to our area another food-reuse app called Olio.[21]

If food is absolutely no longer fit for human consumption, it can be repurposed into animal feed or compost.[22] The potential exists for pigs to consume up to 65 percent of their diet in food wastes. Australian pigs are already chowing down on everything from brewing and distilling wastes to discarded coffee whiteners, Nutella, and peanut butter. In Japan legislation stipulates that animal feed be the preferred use of food wastes (rather than composting or incineration) and that companies that produce food waste purchase pork certified to have been raised on more than 20 percent organic wastes. Housefly larvae and mealworms reared on food waste are also fed to pigs, chickens, and aquaculture-raised fish.[23]

Food waste can be used to produce energy through small-scale farm biogas digesters or large waste-to-energy plants.[24] I first encountered biogas while visiting a small spice farm in southern India. As I remember, it was a simple biodigester that used manure, leaves, and other organic wastes. It was next to a modest concrete home, with pipes carrying the methane produced through anerobic digestion to a kitchen stove. Far away from the Indian spice farm, in London’s Calthorpe Community Garden, another small anaerobic digester turns neighborhood food waste into biogas for heating the greenhouse and cooking in the garden’s café.[25] At the other end of the spectrum are industrial-scale biodigesters, which are not only heating homes and fueling stoves but also powering passenger vehicles, with plans for powering trains and boats in the near future.[26]

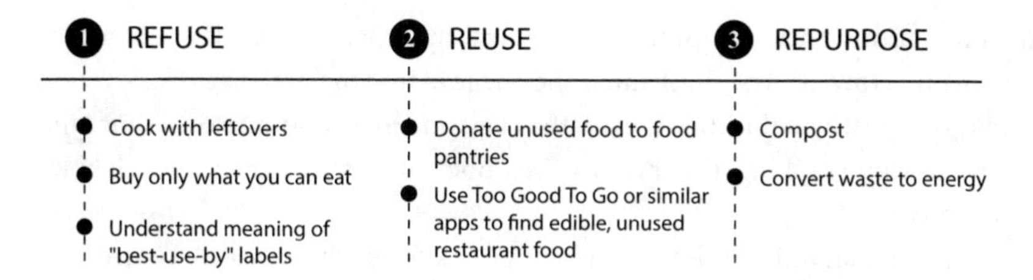

Food-waste hierarchy. Table by Xiaoyi Zhu, redrawn from ReFED, accessed October 23, 2021, https://refed.org/.

Even though people in richer countries waste way too much food in their homes, they are not happy about it. Research by the United Kingdom's Waste and Resources Action Programme revealed that over 70 percent of people are bothered by the costs of wasting food.[27] Food waste also generates visceral reactions: rotting food is disgusting, while discarding perfectly good food that could feed the poor feels wrong to just about everyone who contemplates it.[28] And because no one wants to waste food, it's not obvious who the food-waste bad guys are. We can't blame the fossil fuel behemoths for squandering food. We could turn our wrath on our grocery stores, which are responsible for about 13 percent of food wasted in the United States, except that we might actually like our local grocers and be inclined to work with them rather than against them. And it's important to remind ourselves that nearly 40 percent of food waste in the United States occurs in households.[29]

Food Waste: A Tale of Two Theories

How can we change these wasteful behaviors? Two theories shed light on how we might change what we do with our household food waste. One theory views food waste as an intentional behavior, while the second theory sees wasting food as an unintended consequence of our shopping, cooking, parenting, and other everyday practices.

The Gap between Good Intentions and Bad Behaviors

In 1980 two psychologists, Icek Ajzen and Martin Fishbein, proposed the theory of reasoned action. The psychologists hypothesized that if people *intend* to perform a particular behavior, they are likely to actually perform that behavior. Working backward, a person's positive or negative *attitudes* toward the behavior and how they think others judge the behavior (their *subjective norms*) determine their intentions. So, for example, if someone thinks that smoking is gross (a negative attitude toward smoking) and if they believe that their friends disapprove of smoking (negative subjective norm), they will *intend* to stop smoking. Once they intend to stop smoking, the theory of reasoned action would suggest, they will stop smoking.[30]

Realizing that our best intentions can be thwarted when we lack control over a behavior, Ajzen proposed a second theory, which is still widely used in studies of food waste and hundreds of other behaviors. The theory of planned behavior adds one additional predictor of intentions: *perceived behavioral control*, or a person's beliefs that they have the resources and opportunities to perform a particular behavior.[31] For example, a smoker believes he can access the medications and social support he needs to control his smoking behavior. The theory of planned behavior would predict that because the smoker views quitting smoking positively (attitudes), believes that others approve of quitting (subjective norm), and thinks the behavior is within his reach (perceived behavioral control), then he intends to quit, and he will do it. But we know from experience that even if we believe smoking is under our control, intending to quit doesn't amount to quitting.

How does the theory of planned behavior play out when applied to understanding how people deal with food waste? Unfortunately, the results are mixed. One study showed that people who held negative *attitudes* toward wasting food threw less food out, as one would expect.[32] But another study showed people's negative attitudes toward wasting food had no relationships to their intent to reduce food waste.[33] Both of these studies found that factors not taken into account in the theory of planned behavior, such as habits and number of children in the home, predicted food-waste behaviors. Studies that looked at *subjective norms* in determining people's food-waste behavior also had mixed results, perhaps because what we do with our food waste occurs in our kitchen and so is largely invisible to neighbors who might cast a disparaging glare.[34] *Perceived behavior control* seems to offer more promise in predicting whether people reduce food waste. In multiple studies, people who believed they could control how much food they wasted ended up tossing out less food,[35] although at least one study did not find such connections.[36]

A widely recognized problem with the theory of planned behavior is that most studies stop at intentions rather than measure whether people actually reduce food waste, not taking into account the often yawning gap between our best intentions and our behaviors. Studies that were able to go beyond what people *intended* to do and measure what they actually *did* found that intention to reduce food waste had a strong,[37] moderate,[38] weak,[39] or no relationship to actual food-waste behaviors.[40] In one study, intentions were even negatively correlated with people reducing food waste.[41] In other words, those who intend to reduce food waste may or may not reduce food waste.

No one intends to waste food. Wasting food costs us money, makes us feel guilty, goes against values such as thriftiness, and just seems wrong.[42] But "stuff happens." Like other parents, my husband wants to provide for his children. Whenever we celebrate a birthday or holiday, he offers a sumptuous spread of olives, brie, and humus, and later we struggle to finish the leftovers before they go bad. I once cleaned out my mother-in-law's refrigerator and was shocked to see tofu that was a year past its best-if-used-by date; it was somehow frozen, so it may have been okay, but I was definitely not going to chance it. And last summer our refrigerator was on the blink—we couldn't get it cool enough and kept finding food that normally would have lasted several weeks going bad in a couple of days. So, despite my best intentions, I ended up flinging food into the compost bin. These moments stick in my mind, but perhaps I am not even conscious of how much food I toss out in an ordinary month. In a UK survey, even those who were adamant that they didn't waste food still threw out an average of ninety kilograms of food per year.[43] In the end, maybe we don't even think about our everyday routines of eating and throwing out what's left on the plate.[44] Daily realities can override our good intentions to not waste food.[45]

Motivations, Opportunities, and Abilities

In 2020 the National Academies of Sciences, Engineering, and Medicine published the comprehensive *National Strategy to Reduce Food Waste at the Consumer Level*, which proposed another framework, "Motivations, Opportunities, and Abilities," to understand food-waste behaviors. The framework posits that "consumers are most likely to act in a particular way when they not only are motivated to do so but also have the ability and opportunity to act on that motivation."[46] The report recommends launching a food-waste literacy campaign to increase consumers' motivations and ability to act—for example, by explaining the financial and climate impacts of wasting food and by helping people avoid the temptation of "ten apples for ten dollars"–type specials when grocery shopping. The report also recommends increasing opportunities for consumers to engage in sustainable practices—for example, by implementing citywide free organic waste pickup from residences.[47]

Connecting Food Waste to Household Practices

> **Environmentally unsustainable patterns of consumption have less to do with individual consumers than with the collective development of what we take to be normal ways of life—such as daily showering and meat-heavy diets.**
>
> —David Evans, economist, and colleagues, *Constructing and Mobilizing "the Consumer"*

Given all this conflicting evidence about the relationships between attitudes, norms, behavioral control, intentions, and behavior, perhaps it's time to look elsewhere to explain why we waste so much food—and what we can do to change these climate- and wallet-unfriendly practices. In fact, the theory of planned behavior was intended to predict behaviors of individuals, whereas wasting food is a household social practice, with different family members or apartment mates assuming different roles and routines.[48] Furthermore, our food-waste and eating practices are tied to other household practices, such as cooking.[49]

Behavior versus Practice

In this book, I generally use the term "behavior" except when specifically referring to practice theory, but the distinctions between behavior and "practice" are squishy. What is important is that practice theory zooms out from a narrow focus on what people do—or on individuals and how we can change their behavior—to a broader consideration of the multiple structural factors that influence a behavior or practice, including anything from whom we perform the practice with, to the physical objects or apps we interact act with while performing the practice. Using practice theory, we would not focus on food waste in isolation but rather the interaction of how we treat food waste with other daily practices, such as eating, shopping, parenting, and cooking.[50]

Practice theory offers an alternative to focusing narrowly on what an individual believes, feels, intends, and does. Instead, the focus is on our practices—the routines and the new ways of doing things that play out in our daily lives.[51] Practice theory also switches the focus from reducing the climate footprints of individuals to reducing the climate footprints of their food-waste and other practices.[52] It explores how our daily practices are connected. It also explores how government and business policies influence our practices as well as how our practices can influence government and business policies.[53] Although practice theory is descriptive and thus cannot statistically predict behaviors, it is useful in gaining a broad understanding of why we waste food and what we might do about it.

Imagine an environmental nonprofit is trying to change how I treat food waste in my household. Proponents of the theory of planned behavior might conduct a campaign to change my attitudes toward food waste—to convince me it's a bad thing to do. They might ignore the structural factors, things such as the fact that I may live in an apartment where composting is difficult and my city does not collect food waste, both of which limit my options. In contrast, proponents of practice theory would help me engage in actual food-waste practices, such as cooking tasty meals featuring leftover ingredients.[54] Participation in practices over the long term can give rise not only to new knowledge and skills but also to new values, attitudes, norms, and even identities, all of which can lead to new practices.[55] Thus, helping people be creative with leftovers may have a longer-lasting impact beyond simply changing a cooking and food-waste practice. It could also change how people think of themselves, what they care about, and maybe even whom they spend time with.

Practice Elements

To move our family and friends toward more sustainable practices, it might help to understand what practices are made up of. What makes a practice a practice? Envision a delicate web of what we believe, what we do, and what we can touch or click on, such as a cookbook or recipe app. These practice elements are referred to as meanings, competencies, and stuff.[56]

- *Meanings*, in the context of food waste, are our beliefs and values related to food waste and to what we do with extra food. Maybe we view food waste with disgust or as a resource to be turned into "black gold" (compost), or we believe throwing out food is immoral.
- *Competencies* refer to our knowledge, skills, and routine behaviors or habits. Knowing how to interpret best-if-used-by dates or being skilled at creating delicious meals from leftovers would be counted as competencies.
- *Stuff* refers to physical objects, such as compost bins, leftover recipes, and food. It also includes technologies, such as food-tracking apps used to locate edible food remaining when a restaurant closes that is sold for a low price to anyone who shows up.

The web of meanings, competencies, and stuff (objects and technologies) is fragile; if one of these elements changes, the practice—whether sustainable or unsustainable—unravels.[57] As I researched the problem of food waste, I discovered that carrot tops—the feathery leaves that get chopped off at the store or when I am ready to eat a carrot from my garden—are edible. I even

"It says 'Best by 47 BC,' but it smells fine!"

came across carrot greens recipes, such as cashew carrot-top pesto and carrot-top tabouli salad, on the Oh My Veggies website.[58] So, I can now readily cook with carrot tops. This new cooking skill has certainly influenced whether I toss the tops into the compost bin when I harvest a carrot from my garden. Once my thinking about carrot tops changed from food waste to food resource and I mastered cooking with carrot tops, my previous practice of throwing out carrot tops—wasting food—was transformed into a new practice of eating carrot tops and throwing less food into the bin.[59] My carrot-top web of meanings, cooking competencies, and available recipes has changed, and—presto!—my carrot-top waste practices have been transformed.

At a higher level, related practices influence each other and thus form a web of their own; cooking, shopping, wasting food, and raising children are interwoven in such a web.[60] Shopping practices that impact how we treat our waste include frequently buying more than we need or can eat or impulsively grabbing items being promoted with "buy-one-get-one-free" deals.[61] Planning practices that minimize food waste include checking the food we have on hand and making a list of what we need prior to shopping as well as sketching out meals in advance.[62] Once food

What Exactly Does "Best Use By" Mean?

According to the US Food and Drug Administration (FDA), Americans mistakenly toss out $32 billion worth of food each year, just because they are misreading date labels. Consumers assume the best-if-used-by labels indicate food safety, when in reality they are an estimate of the last date when food maintains peak quality and flavor. Adding to the confusion is that the wording of labels varies: you may have seen "sell by" and "best by" on food items. With the exception of labels for baby food, no date labels are regulated in the United States, although the FDA is working to establish guidelines in order to reduce food waste.

In response to the question "Can I eat a food after the expiration date?," the FDA website states emphatically, "Yes! This is because it's not an expiration date. A food which doesn't show signs of spoilage after the specified date can still be eaten. Remember that food manufacturers have an incentive for you to only consume products in their 'peak quality' because then you will buy their products more often."[63]

is purchased, people's cooking practices, such as how creative they are with leftovers, impact how much food they toss out.[64] Changing our living situation can also impact how much we waste. When we get new roommates, when our children move out, or when we retire—any of these transitions might be propitious times to try changing practices such as what we eat and what we waste.[65]

Transforming Household Food-Waste Practices

If we want to change our household food-waste practices, the trick is to home in on what elements of wasting food and related practices we might be able to alter.[66] The possibilities are vast for the meaning element of wasting food because there is much room to transform what we view as edible (for instance, apple and squash skins or soy milk whose best-if-used-by date has passed). For competencies, options include learning to cook new dishes with leftovers and use so-called unavoidable waste, to shop with forethought, and to store perishable foods properly. In the stuff category, improvements are often spurred by new technologies such as a food-waste app or by new items such as a cookbook for being creative with leftovers.[67] This is especially the case when new technologies and new meanings or ideas seem "cool"—they generate positive emotional energy, so people want to join in the new practice.[68]

For food that is no longer edible, food-waste bins are a stuff element that enables food to be repurposed. If you live in Seoul, South Korea, you likely would not be putting food waste in the garbage because it's illegal to do so. Instead, city dwellers take their food waste to "e-bins," often located in apartment parking areas, which weigh and track their banana peels, eggshells, and remnants of their takeout meals. The bin then levies a charge, according to how much waste is dropped off. The city in turn uses the food waste as animal feed and to generate compost and biofuel.[69] Some Swedish and North American city governments, as well as nonprofit groups, also provide kitchen food-waste containers that residents place on the curb for pickup or drop off at a collection site.[70] Practice theory would predict that by changing stuff—that is, by providing a food-waste bin or new app—a city can transform its citizens' and businesses' food-waste practices.[71] In any campaign based on changing consumer practices, attention to the stuff element cannot be overemphasized because of the potential of physical objects and phone apps to expand what people are readily able to do.[72] One study showed that providing households with a compost bin, a refrigerator thermometer, and information, such as recipe ideas to create new meanings around "aging" food, yielded a 78 percent reduction in food waste.[73]

What's more, because different practices are interwoven, altering one practice may shift another. If we start to make a shopping list rather than buy impulsively, we have changed our shopping practices, which in turn impacts how much food we waste.[74] Just last week, I went shopping for ingredients to make granola. I don't like making granola, so I try to make a large batch that will last a couple years. I decided to buy my supplies on a day where the local co-op had a "20 percent off" special. To make a long story short, I ended up buying twenty-five pounds of oats, twenty-five pounds of sunflower seeds, and twenty-five pounds of pumpkin seeds, plus five large bags of flaked coconut, a good amount of almonds, and other staples such as vitamins and large tubs of brewer's yeast and cashew butter. When I got to the checkout, the clerk informed me that the deal was 20 percent off but with a maximum total discount of twenty dollars. Once I got home, I panicked, realizing I had gone wild spending hundreds of dollars on supplies to save

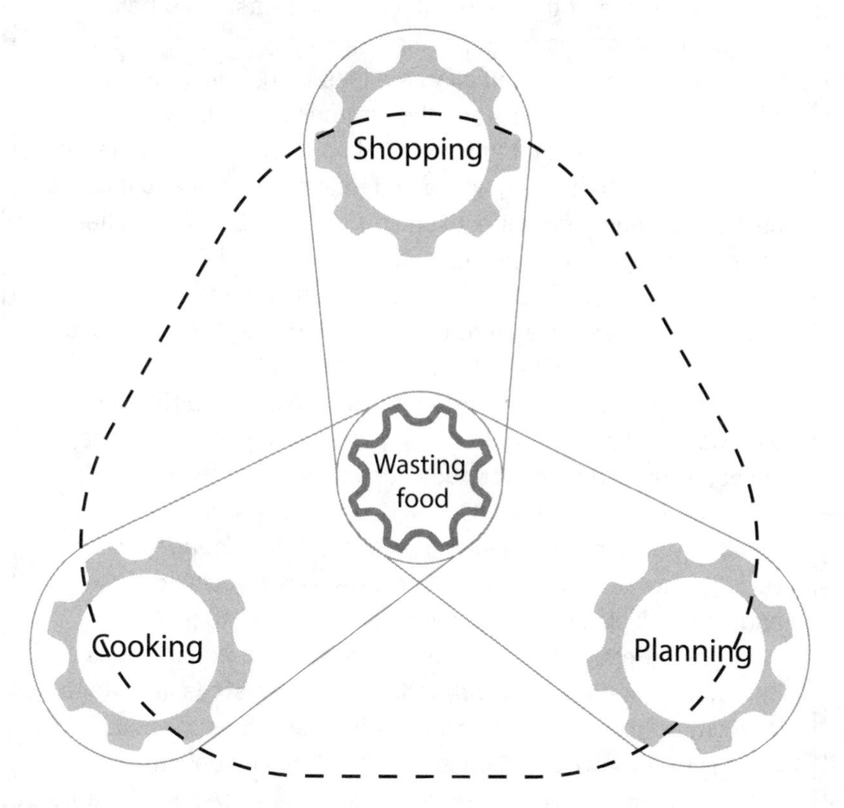

What we do with our food waste is connected to how we shop, cook, and plan for meals and food purchases. (Parenting and other family practices are also connected to food-waste practices.) Diagram by Xiaoyi Zhu.

Love Food Hate Waste

Reducing food waste may not seem to warrant a lot of fanfare. But the launch of the UK's Love Food Hate Waste (LFHW) campaign in 2007 took place at London's renowned, one-thousand-year-old Borough Market, one of the city's top ten tourist attractions.[75] Actor, singer, and celebrity chef Ainsley Harriott demonstrated food-waste-saving recipes and tips as the nation's environment minister stood by.[76] Over the next six years, the UK reduced food waste by 15 percent and saved consumers a total of £3.3 billion a year.[77] This was definitely something for the country to celebrate and for others to emulate.

According to its website, Love Food Hate Waste "aims to raise awareness of the need to reduce food waste and help us take action. It shows that by doing some easy practical everyday things in the home we can all waste less food, which will ultimately benefit our purses and the environment too."

Informed by research that revealed how consumers in the UK think and feel about food waste and reflecting ideas about connections among different household practices, LFHW identifies three simple practices that can reduce household food waste:

Purchasing: Households buy the right amount of food.

Storing: Households store their food to preserve its shelf life.

Eating: Households eat food before it goes bad.

To help households master these practices, the LFHW website features tips for how to "keep more of your humble spuds, crisps, chips and other potatoes goodness out of the bin," "discover the secrets of food planning," and "make your freezer your friend." There's a "compleat recipe book," full of ideas for converting leftovers—such as stale bread—to delicious side dishes, and short articles with such titles as "5 Little Wins to Save Money and Reduce Food Waste."[78] Whereas the website does include information about why reducing food waste is important, its focus is not on trying to change attitudes—most people already believe wasting food is bad.[79] Rather, the upbeat and colorful website provides consumers with the know-how, the emotional energy, and a portion calculator to help them save food and money.[80] In the United States, groups such as Save the Food and StillTasty have launched similar food-waste campaigns.[81]

LFHW partners with over three hundred local governments to get practical advice out to their constituencies. Working with schools, LFHW has shrunk plate sizes to reduce food waste in cafeterias and supported student-designed projects such as the "Be a Winner, Not a Binner" campaign and a leftover-soup-making competition.[82] By 2013, seven years after the launch of LFHW, private food businesses had invested over £15 million in food-waste reduction communications to consumers, exceeding the government's funding for the campaign.[83]

twenty dollars and became concerned whether we would ever eat all the granola before it went bad.

If we strive to spend as little time shopping as possible, we likely fill our grocery cart with every conceivable item our family might need for the next several weeks, some of which may never be consumed. Alternatively, if we are fortunate enough to have a fresh-food market that we can stop by every day on the way home from work, we buy small amounts, and fewer items spoil before we get to them. If we cook for a family, especially one with small children, the health of our children is utmost in our mind. We are super attentive to expiration dates and throw out soy milk immediately upon reaching the best-if-used-by date without bothering to check for mold or a putrid odor. And some of our dishes will be rejected by children, maybe even tossed from the high chair onto the floor—thus contributing to our family's food waste. All these connections among shopping, planning, cooking, parenting, and food-waste practices add to the complexity of influencing how much food ends up in the bin.[84]

Food Rescue

> **The image of people who can afford to buy food but instead choose to get it from the garbage, and are living really well off that food—that's an incredibly compelling image, and the freegans in New York were very savvy in using that.**
>
> —H. Claire Brown, journalist, *Freeganism*

Walking down the sidewalk in Manhattan means dodging mountains of black plastic bags filled with putrescent refuse, set out each day in front of stores and restaurants. Without City Harvest, the trek would be even more treacherous and, frankly, more malodorous. Inspired by ancient gleaners, who collected food remaining in the field after harvest and distributed it to needy families, today's city harvesters use refrigerated trucks to round up nearly three hundred thousand pounds of food daily, or over one hundred million pounds of food "surplus" in a year. Once "gleaned" from restaurants and stores, urban farms, and food manufacturers, the surplus is given to food pantries, soup kitchens, and community food programs, where it helps feed over a million New Yorkers who struggle to put food on their tables.[85] The Food Rescue Locator has a database of similar food collection groups across the United States, with such names as Aloha Harvest (in Maui), Food for Free (in Cambridge, Massachusetts), Plenty (in Floyd, Virginia), and Friendship Donations Network (in Ithaca, New York).[86] Complementing these efforts, the National Gleaning

"I like it when nobody orders the fish."

Project compiles legal and policy resources to support the work of food recovery organizations.[87]

Food surplus not donated by a store or restaurant winds up in the dumpster, where it might be fished out by a dumpster diver. Dumpster divers eat the food themselves or share it with friends or strangers, often with an invitation to join in their food-rescue practice. A co-op in Eugene, Oregon, would routinely put its edible food waste outside the store, inviting people in need to be the first to recover the items. They were followed by the more affluent political dumpster divers who, while not food insecure themselves, shout out their condemnation of rampant consumption and waste.[88]

In 2015 several activists decided to take over an unoccupied storehouse underneath Paris's Boulevard Périphérique (belt highway). They converted the

Dumpster Diving in Ithaca

A reviewer of this book recommended that I should try dumpster diving myself instead of just writing about it. Since that reviewer was my adult son, I invited him to come along, and then my younger son decided to join the midnight adventure. We first pulled up to the dark, deserted parking lot of an Aldi supermarket and discovered that its dumpster seemed to be in a locked garage. So, we went on to our local GreenStar co-op and found a row of compost bins. Score! We gathered bags of lettuce—seemingly the outside leaves that were chopped off lettuce heads before they were placed on the shelves—as well as a blueberry muffin, part of a carrot, an apple with a small bruise, and three perfect string beans. Next we hit the Asian grocery market, whose dumpster nearly knocked us out with its rotting fish and Chinese cabbage, but we hit pay dirt with four blueberry donuts—which it turned out were not Asian at all but, according to the sticker on top of the plastic container, had been purchased at Tops grocery store. After a disappointing stop at Wegmans, which seemed to have all its waste go directly from store bays into closed trucking containers, we hit Tops. Tops had a similar system of waste going directly into a container—but one side of the container had a slit. When my younger son put his hand through the slit, it was like a slot machine—out poured a package of Bada Bean Bada Boom, a box of After Eight chocolate mints, and two bags of No-Bake Cookie Company dark chocolate, peppermint handcrafted cookies—all past their expiration dates but which turned out to be quite tasty—as well as three cans of alcoholic seltzer water (mango, grapefruit, and peach flavors). Our penultimate stop at Trader Joe's revealed dumpsters behind an easily scalable fence, but we decided not to risk arrest. And at our final stop at Ithaca Bakery, we opened a gate to what we thought might be the dumpsters and were confronted by an employee taking a smoke break who, somewhat taken aback, asked, "Do you need anything?"

Once we got home, we laid our haul out on the kitchen counter and sampled the goodies. The donuts were fresh, the muffin stale, the carrot and apple good, and the lettuce bitter. I enjoyed the Bada Bean Bada Boom snack, although my older son was disgusted by its dog-food-like appearance and spit it out.

We realize that for us what was an adventure could be a matter of life or death for our hungry brethren. Our conclusion, based on a single night's sampling, was that if you had a car, you could go from store to store (perhaps a three-mile distance in Ithaca) and maybe get enough to eat for the day—but it would not be nutritionally balanced.

We also realized how important expiration dates are. Who knows how many packages of Bada Bean Bada Boom snacks were filling the dumpster container—expired yet perfectly edible if you like fava beans with onions and garlic. Since our midnight adventure, I have volunteered for the Friendship Donations Network and seen that if bread is even a day past its expiration date, it gets sent to hungry livestock instead of being donated to hungry people.

Upon sharing our adventure with a friend in Prescott, Arizona, I learned that she had been an avid dumpster diver in college, feeding herself and her roommates over four years. But she acknowledged that that would be nearly impossible today because many stores, like some of the ones we encountered, move their expired food directly from the store into a huge container or have closed off their dumpsters with locked gates.

However, all is not lost. It appears that food-donation programs are growing and more food is going directly from stores and restaurants to food pantries and other locations that feed poorer residents. I recently returned to Wegmans, this time as a volunteer with the Friendship Donations Network. Together with two other volunteers, I sorted and weighed boxes upon boxes of bread, fruit, and prepared meals that Wegmans had laid out for us on the loading dock. We packed our cars and whisked the food away to the network's food-distribution center. There we helped stock the shelves while a "shopper" gathered free food for her family.

abandoned building into the Freegan Pony Restaurant, thus "freeing" the ghosts of the horses that had once been slaughtered at the site. Instead of serving meat, volunteers would rescue fresh produce from nearby wholesale markets, where normally one rotten tomato means the entire crate is tossed out. Professional chefs created the *menu du jour* using whatever was recovered that morning, and customers paid what they could afford for their vegetarian meals. Within a year, the Freegan Pony had become what one review called the "coolest and hippest spot in Paris."[89] It also had become a battleground, pitting authorities obligated to shut down a restaurant squatting on city property against satisfied diners and activists who supported the mayor's commitment to fight food waste.[90]

Dumpster divers, gleaners, and the Freegan Pony are all part of a social movement known as freeganism (an amalgam of "free" and "vegan"). Adherents of freeganism share resources and reduce and recover not only food but all things discarded. The suite of provocative freegan rescue-and-reuse practices that have emerged around grocery store and restaurant dumpsters reframe food waste as "food surplus" that feeds neighbors in need. Regardless of whether they are rescuing day-old donuts or a crate of cantaloupes, freegans share an antipathy toward "the social and ecological costs of an economic model where profit is valued over the environment and human and animal rights."[91] Their practices challenge mainstream consumption and capitalist values, while helping to create a pathway for food from dumpsters and bins to countertops and compost.[92]

The words of the ethicist Eric Godoy capture how dumpster diving and other freegan practices are not just ways to reduce food waste but also strategies to attract others to the antiwaste cause: "Structural change can be enacted by governments through new laws and policies, but it also requires imagining and experimenting with new ways of thinking and living, as well as attracting a critical mass of supporters for those innovations. Engaging in these practices should be thought of as a political act, and an empowering one at that since, although it can inspire government action, it does not rely upon it."[93]

Vote with Your Fork

> **In addition to openly political forms of action (such as boycotts or political mobilization), consumers through their everyday practices, consciously or unconsciously, leave an active mark on these larger social systems.**
>
> —Frank Trentmann, historian, *Citizenship and Consumption*

In Germany dumpster divers publicly scorned the grocery stores and their wasteful practices on Facebook: "Some retailers really suck. They keep the bread inside the store until it perishes, so that no one can eat it anymore. Such idiots! Others try to fight back with fences. And others put disgusting stuff or coffee grounds on top of the good stuff so that you do not find the good stuff. Or it becomes so disgusting that you don't want to lay your hands on the nice stuff."[94]

The dumpster divers' angry rhetoric provoked the grocers to have them arrested for trespassing. The legal proceedings became news, generating sympathy for the dumpster divers in their David-and-Goliath fight against grocers cast as the cruel, money-grubbing food behemoths. Media coverage of the powerful food industry bringing charges against a couple of scraggly dumpster divers laid bare the contradictions in the food system for all to see. How could the grocery stores throw out so much good food when people were hungry and food production created such enormous inequities—not to mention colossal climate and environmental footprints? The charges against the dumpster divers were eventually dismissed. By then it had begun to dawn on the grocery stores that they needed to improve their image, even if they were not going to support the anticapitalist message of the dumpster divers and other freegans.[95]

Regardless of whether you agree with their politics, you have to admit that the German dumpster divers and French freegans are actively participating in food systems and democracy. "Voting with your fork" captures the idea of being a productive, nonpassive citizen through consumer practices such as rescuing

GoZeroWaste

The Beijing financial analyst Elsa Tang launched GoZeroWaste in 2016 after becoming appalled by the waste piling up around her. She conducted her own zero-waste life experiment and managed to fit all her waste over two weeks into a small jar. To spread her new lifestyle to others, she launched a WeChat social media group and challenged followers to implement one new waste-reduction activity each day for twenty-one days. Today GoZeroWaste has twenty-eight WeChat groups in different cities, which promote in-person events such as swap days where people exchange used items. It is likely the biggest zero-waste community in China and is committed to making zero waste a lifestyle trend.

In addition to zero-waste behaviors spreading from person to person, GoZeroWaste's behavioral challenges are spilling over to new types of action. My Cornell students worked with GoZero-Waste to design a plant-rich-diet seven-day challenge. GoZero-Waste's work also has spilled over to encompass consulting and conducting workshops for schools and businesses. As of spring 2021, it had conducted over four hundred such events in twenty cities across China.

Note that in China, where registering a nonprofit is not easy, GoZeroWaste demonstrates an alternative organizational form—social media groups to promote sustainable practices. The groups sometimes morph into social enterprises (businesses formed around a public good), and in fact GoZeroWaste is registered as a company in Beijing. As GoZeroWaste gains notoriety and followers, its influence on businesses and government policy may grow. Although China is not a representative democracy, its corporate executives and government officials are concerned about the issues people care about and respond to pressure from below. Undoubtedly, when a video featuring a young couple living a zero-waste lifestyle in Beijing is viewed by thirty-eight million people, government officials start to realize that China's younger generation wants to do something about rampant waste.[96]

food from dumpsters to feed the poor.[97] Through our choices about what we eat, where we get our food, and how we handle food waste, we can demonstrate that we are knowledgeable and politically active—that we are good food citizens supporting a just and environmentally sustainable food system. Good food citizens engage in direct actions, such as volunteering with a food-rescue organization to transport unused food to soup kitchens or supporting local farmers through buying at farmers markets. They also conduct indirect actions, such as advocating for city-wide organic waste pickup or statewide food-waste laws.[98]

Because we eat every day, we can vote with our fork every day. Voting with our fork is a way to bypass the political system but still be heard—it transcends political jurisdictions and operates directly between the "voter" and private businesses, such as farmers markets and grocery stores.[99] It is important to note, however, that not everyone has access to or money to pay for local organic veggies, food-waste collection services, and other components of food citizenship. This means that the middle class and wealthy can participate more readily in ethical food citizenships practices; they have more "votes" about what our food systems should look like relative to poorer people.[100] Yet in the United States and other countries, low-income people have historically used consumer practices as a means of social change—from the restaurant sit-ins during the civil rights movement to the grape boycotts of the farmworkers movement led by Cesar Chavez. This suggests that resources apart from income per se, such as social capital, expertise, and strong faith-based institutions, enable "voting with your fork."[101]

Consumer Social Movements

> **Sometimes they use the purchases of "ethical" products like signatures on a petition.**
>
> —Clive Barnett, geographer, *Globalizing Responsibility*

Composting or eating vegan may be seen as small individual environmental behaviors—a way for people daily faced with dilemmas around "doing the right thing" to take what feels like meaningful action.[102] But how can these individual household actions become collective and widespread and thus meaningful in the face of the colossal climate crisis?

In one sense, these seemingly individual practices are collective—they are performed by hundreds or thousands of people influenced by organizational campaigns. Nonprofits such as Love Food Hate Waste and startups such as Too Good To Go engage individuals across national borders in reducing food waste. They often do so by approaching people not primarily as consumers but through

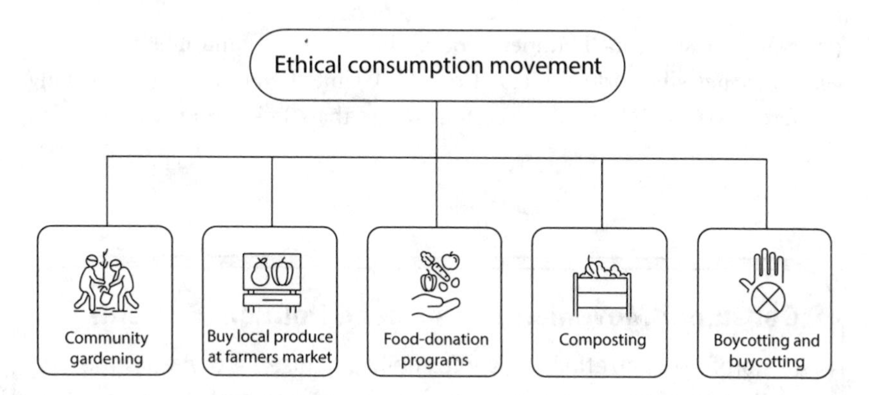

Ethical consumption integrates multiple food-related practices. Chart by Xiaoyi Zhu.

their more salient identities, such as parents or members of a faith group or profession. (After all, who thinks of themselves primarily as a consumer rather than as a mother, a Muslim, or a high-tech worker?) This enables the organizations to recruit not just individuals who want to save the planet but also people with diverse motivations and identities (e.g., mothers who see reducing food waste as a way to save money). Furthermore, it enables nonprofits to recruit entire networks, such as all members of a church or all employees at a socially minded enterprise.[103] As more and more people take part in these and related consumer practices, we end up with a sort of "green contagion" spurred on by the emotional energy of doing something fun, social, and meaningful.[104]

Yet these nonprofits, informal groups, and startups have a larger mission than simply the spread of ethical practices among thousands of consumers. They form the backbone of consumer social movements, such as ethical consumption and political consumerism,[105] which use consumer practices to transform industrial food systems into something more just and sustainable. Boycotts, food rescue and donation, and buying local are forms of protest that can be organized by these social movement organizations and that individuals can participate in daily and through familiar venues such as workplaces, community groups, and restaurants. Nonprofit organizations that institute fair trade certification programs for food products, dining hall food-waste pickups for donation, animal feed, or composting, and healthy food procurement policies in zoos, schools, and churches transform the ordinary places where people go to shop, learn, and worship into "advertisements" for sustainable consumption.

In this "empowered" view of consumerism, or so-called life politics,[106] consumer practices become political acts.[107] They are combined with other types of protest. Greta Thunberg doesn't just eat vegan and take the train to Glasgow. She tells world leaders at the 2021 United Nations Climate Change Conference (COP26) in no uncertain terms that their promises are simply "blah blah blah."[108]

Consumer Movements Transcend Political Systems

Nonprofit organizations and informally organized groups are the backbone of consumer movements. In China Operation Empty Plate sponsored a grassroot campaign on the Twitter-like app Weibo. The campaign was featured on the front page of the *People's Daily News* after it gained the attention and support of the Communist Party leader, Xi Jinping.[109] Later President Xi launched the Clean Plate Campaign, calling the amount of food wasted in China "shocking and distressing."[110] And in the United States, the nonprofit ReFED hosts a food-waste "policy finder" webpage so citizens can track and offer their support for proposed legislation covering everything from developing a food expiration date labeling system to food-donation programs.[111] Even the adversarial dumpster divers in Germany, after a series of legal challenges brought by grocery retailers, forged common ground with over three thousand stores. Rather than toss their slightly blemished apples or day-old donuts into the bin, the stores agreed to hand over their waste to the organized activists for redistribution to the hungry.[112]

Regardless of whether the political system allows protest demonstrations or even voting, people can engage in sustainable consumer practices. And because food and other production systems are global in nature, consumer activism, such as "buycotting" (buying more) fair trade products, can transcend national borders. In this way, consumer movements become an effective form of political activism in a globalized economy.[113]

Try and Try Again

We started this chapter with the story of Mark Schlesinger talking about how he and his wife, Fran, who were attempting to convince their neighbors that they could compost their food wastes with a minimum of stress. Since then, Schlesinger has become ever more active trying to institute structural change in his community. When I touched base with him last fall about how things were progressing, he wrote:

> I wish I could tell you that good things have happened quickly. It's not the case. My local efforts to implement and expand composting at Duxbury Estates, our condo development with 40 units in duplexes and free-standing single-family homes, have met with interest and questions, but so far Fran and I are the only residents using the services of Black Earth, which picks up composting weekly at the rate of $18/month. (They offer a bag in return every year.) The last time I inquired at a meeting of Sustainable Duxbury, a couple months ago, efforts to engage the local transfer station had not progressed. As a relatively new member, unknown in the town, I'm hesitant to take up that charge, but I continue to urge it. In addition, I've engaged a member of our 5-person Select Board to raise with the Board the matter of how the town can facilitate composting, either by implementing it at the transfer station or other means. Finally, I'm preparing a letter to our condominium community describing our progress with Black Earth, with which we are well pleased.

Schlesinger 's experience illustrates some of the challenges in moving from relatively stress-free individual behaviors such as composting to advocating for systemic change. Even working with a local organization—Sustainable Duxbury—in his small city, his efforts are proceeding slowly. Yet Schlesinger is not giving up. He continues to try to "fan the small flame"—the flame that began with his home composting and that he hopes will gain oxygen as he keeps up his efforts to influence local policies. Schlesinger sent another update a few weeks later:

> This Sunday I'll be one of four speakers at First Parish Duxbury's service. Our purpose is to issue a call to action. I'll talk about composting efforts so far—the reasons to do it, especially—and the notion of "paying it forward," as well as the look in my children's eyes, especially Annie's, when we would just sit on a cliff at West Quoddy Head and wait for something to come to us, a minke, a bald eagle, seals, harbor porpoises, ospreys . . . and the nature that is in us.

After talking with my students about how much food is wasted in the dining halls and fraternities, like Schlesinger I am exploring how Cornell can ramp up its food-waste reduction policies. I have already been told about how difficult this can be given the health regulations around donated food, and my emails to people in charge have gone unanswered. I only hope that I can follow Schlesinger 's example and not give up when encountering the inevitable barriers along the way.

• • •

GLEAN: Don'ts and Dos

1. Don't think of what you do with food waste in isolation from your other food practices. Instead, consider checking what's in the refrigerator before shopping, making a shopping list, making a meal out of leftovers, and using good food that is beyond its best-if-used-by date.
2. Do support food recovery or donation groups. Groups that rescue unsold or unserved food for soup kitchens or food pantries are a "double whammy." They keep food out of the landfill and in the mouths of people.
3. Do set up a compost system on your balcony or in your yard or avail yourself of pickup or drop-off services provided by local government or a nonprofit.
4. Do "vote with your fork" as often as possible. Invite your family and friends along on a trip to the farmers market, to meet with a city council member to advocate for curbside pickup of food waste, or, if they are adventurous, a dumpster-diving expedition.

3

GIVE

In 1950 the average size of a new home in the United States was 983 square feet. By 2015 the typical new house sprawled over 2,687 square feet. Yet the average number of people you would find living in a home declined from just less than three and a half to about two and a half during that same time period.[1] Why were Americans buying larger and larger homes for smaller and smaller families?

The answer lies in the title of the economist Robert Frank's recent book *Under the Influence: Putting Peer Pressure to Work*.[2] Frank believes that what we want—or believe we need—is relative to what our neighbors have.[3] So, as neighbors buy larger homes, we can't help but be under their influence and buy larger homes ourselves. Frank's "under the influence" explanation flips our network climate action argument from chapter 1 on its head. Just as we influence friends and neighbors to install solar panels on their roofs, we can influence those around us, in a climate-negative way, to buy a bigger house or car or even hold a fancier destination wedding. And, unfortunately, luxury buying has an outsized impact on one's carbon footprint. Market research firm Ipsos even coined the term "affluencer" to capture the affluent's outsized influence on what luxury purchases society deems acceptable.[4]

Frank discounts the possibility that our larger homes are making us happier. His views are supported by the fact that over the past three decades as the rich were getting richer, the incidence of social anxiety disorder increased from 2 to 12 percent of the US population.[5] While there may be an initial spike in happiness following buying a large home or car, that spike flattens out quickly. Likely we would be just as happy with small homes but only if our neighbors and friends also reduced the size of their abodes.[6]

In fact, a high level of income inequality, such as found in the United States, can become an assault on our mental well-being. It increases our anxiety as we constantly compare ourselves to those who are richer and as we sink into the ranks of the "ego-insecure." It harms our ability to work together to protect public goods, ranging from neighborhood parks to a stable climate. And the impacts are felt not just by those at the bottom of the income ladder but by all of us.[7] What actually influences happiness more than wealth or how we spend our money are societal trust, generosity, and equality—the very same factors that help societies maintain our public goods such as clean air, public safety, and green spaces.[8]

If everyone would be just as happy if they and their neighbors spent less on houses, cars, and weddings, where would that money go? Frank makes an argument for spending it on public goods. Unfortunately, many governments have not heeded Frank's call to stem the growth of inequality and allocate adequate funds for health, education, natural areas, and averting the climate crisis.

When governments fail to take action on meat consumption, food waste, or women's education, citizen groups try to fill the gap. Similarly, if tax rates are insufficient to thwart oversized wealth accumulation or to pay for climate solutions, citizens will find a way to make some sort of progress nonetheless. Although grassroots protest and other forms of activism are not a substitute for legislation, they can be effective tools for pressuring governments to take action.[9] For these reasons, contributing to nonprofit organizations that fight against inequities and work to conserve public goods is important. It is also a means to support climate solutions such as health and education that don't lend themselves to direct or lifestyle action.

Growth or Degrowth?

Some claim the climate crisis will only be solved by developing better carbon-capture methods, better batteries, and other new technologies.[10] Others claim that we can solve the climate crisis only through expanding prosperity to include those left out of our current system—for instance, by retraining coal and chemical workers.[11] Still others claim that we must abandon capitalism altogether in favor of "degrowth," or a radical redistribution and reduction in the size of the economy. With degrowth, social and ecological well-being is prioritized over corporate profits, overproduction, and excess consumption.[12]

Project Drawdown Solutions Lending Themselves to Philanthropy

Note that although the term "philanthropy" often brings to mind obscenely wealthy philanthropists such as Bill Gates and Jeff Bezos, anyone who donates any amount of money, however small, to a community or nonprofit organization is engaged in philanthropy. The word literally means "love for humankind," and it has come to signify the generosity that stems from that love. That generosity can be directed toward the many nonprofit organizations trying to alleviate climate catastrophe and wealth inequality. No matter which nonprofit website you land on, your eyes likely will be drawn to the "Donate" button, often in a bright color on the upper righthand side of the screen.

Philanthropy is important because it enables us to support organizations fighting for climate justice as well as those focused on specific climate solutions. These organizations include ReFED, the Good Food Institute, and others fighting for policies regarding food waste, lower-carbon diets, and other direct actions. Philanthropy also encompasses organizations such as Femme International that address climate solutions that do not lend themselves to direct action. "Family planning and education," by suggesting limits on human population growth, is a top Project Drawdown solution, albeit one fraught with controversy. Project Drawdown's "land sink" solutions, such as "forest and peatland restoration" and "perennial staple crops," involve capturing and storing carbon in plants and soils. These are examples of the many climate solutions where donations to health and conservation organizations rather than household practices play a role.

Health and Education

Some scientists have claimed that lowering birthrates is the most important climate action a country can take.[13] Yet having children is a highly personal decision, and for many, children bring unlimited joy and even status over a lifetime. Perhaps more than other climate solutions, a focus on health and education—which is closely connected to family planning and population size—can appear insensitive and patronizing at best and downright racist, classist, and colonialist at worst. It's not a good look for people to be telling other people not to have children, and who tells whom can be problematic as well.

For starters, the climate impact of a child born in a rich country is magnitudes greater than that of a child born in a developing country.[14] In 2019 countrywide averages of tons of CO_2 emissions per capita ranged from 0.029 in Burundi to 8.123 in China, 15.519 in the United States, and 38.823 in Qatar.[15] This means that it would take about 1,340 Burundian children to do as much damage to the climate as a single child born in Qatar. Although these figures are based on industrial and other sectors rather than measuring each individual's carbon footprint and although the report does not mention stronger greenhouse gases such as methane, the figures do provide an indication of the deep disparity in emissions and income among nations. The numbers also suggest that the emissions reduction associated with having one fewer child differs exponentially across different countries.

Additionally, many people in developing countries, and poor people in rich countries, have been subject to colonialist and racist practices, including forced sterilization of women. Such atrocities have been documented in articles such as "Perpetuating Neo-Colonialism through Population Control: South Africa and the United States"[16] and captured in terms such as "reproductive imperialism"[17] and calls for "reproductive justice."[18]

Standing behind such colonialist and racist policies were prominent conservation scientists who argued for limiting the populations of nonwhite and differently abled peoples. Louis Agassiz, a biologist and geologist who founded the Museum of Comparative Zoology at Harvard in 1859, promoted "polygenism," a theory that considered different races to be different species. He went to lengths to demonstrate that the Negro race was inferior to the Caucasian race, proclaiming that "the brain of the Negro is that of the imperfect brain of a seven month's infant in the womb of a White."[19] His protégé, the ichthyologist and first president of Stanford University David Starr Jordan, took Agassiz's theories a step further when he became chairman of the Commission on Eugenics in 1910. He was able to secure an endowment for a eugenics laboratory in Cold Spring Harbor, New York, where scientists carried out studies on "feeble-mindedness . . . and the origin and maintenance of superior strains."[20] Although polygenism and eugenics were abandoned, racist ideas persisted among some conservationists and ecologists, including Garrett Hardin and Paul Ehrlich.[21]

Yet how do we weigh these racist currents in environmental conservation with the fact that not providing women with access to reproductive health services and education is itself an injustice? Many poor girls want access to menstrual pads so they can continue their schooling,[22] and access to reproductive health services is associated with empowerment and educational and job opportunities for women.[23]

Recognizing the need to address reproduction-related injustices, Project Drawdown states, "Honoring the dignity of women and children through family planning is not about governments forcing the birth rate down (or up, through natalist policies). Nor is it about those in rich countries, where emissions are highest, telling people elsewhere to stop having children. When family planning focuses on healthcare provision and meeting women's expressed needs, empowerment, equality, and well-being are the result; the benefits to the planet are side effects."[24]

According to Drawdown Lift and reports from the Malala Fund and the Brookings Institution, women have fundamental rights to education, health care, and decision-making about family size and timing.[25] If granted these rights, women will be better able to participate in government and local economies, including the emerging green sector. Women's employment in turn will reduce poverty and birth rates. Because high rates of socioeconomic inequity within a society are associated with environmental destruction, greater equity will also lead to better environmental outcomes.[26] Further improvements will occur because women in government are more likely than men to support environmental policies and sound management.[27] One strategy to enable women to achieve their rights to education and control over family size is for health services to provide free menstrual supplies and contraceptives, thus enabling girls to stay in school and in the work force.

Given that health and education, when equity concerns and racist and colonialist historical legacies are addressed, are critical to the climate and to justice, how does one support this climate solution? I live in Ithaca, New York, which has good schools and a good hospital and where most women have access to birth control and menstrual supplies. Although volunteering with Planned Parenthood or another organization is an option, I already volunteer with several climate groups, which I enjoy and where my skills are put to good use.

What I can do is donate money to Planned Parenthood or another women's health or education organization and try to persuade my family, friends, and colleagues to do the same. This is exactly what Annalisse Eclipse, an undergraduate student in my Climate Solutions course, did. She chose the organization Femme International, which empowers women and enables girls to stay in school "every day of the month" by providing access to sustainable menstrual products and shattering the stigma surrounding menstruation.[28] Eclipse invited friends and faculty to a showing of the Netflix documentary *Period: End of Sentence*, which follows a group of women in rural India who launch a low-cost sanitary pad movement. After the movie ended, Eclipse offered us some snacks and invited us to contribute to her fundraising campaign for Femme International. Several of us contributed before leaving the room. Eclipse showed how by bringing people together for a fun activity (and offering the gift of food), her philanthropic behaviors spread to others.

Land Sinks

In addition to being the lowest greenhouse gas emitter per capita of any country worldwide,[29] Burundi is also one of the poorest nations in the world, ranked as third poorest globally in 2017.[30] Yet Burundian climate fellow Emmanuel Niyoyabikoze launched the Greening Burundi Project, an NGO that engages school children and volunteers across his country in planting trees.[31] Similar grassroots tree-planting groups have sprouted up all over the world,[32] as have massive global initiatives such as the Nature Conservancy's Plant a Billion Trees campaign.[33] The many tree-planting and tree-care local nonprofits and international NGOs offer ample opportunity for philanthropy.

Tree planting would fall under the Project Drawdown solutions category "land sinks," which refers to the fact that land is made of plants, trees, and soil, all of which are rich in carbon that they take up from the atmosphere. More trees and more soils left intact mean more carbon is sucked out of the atmosphere. Project Drawdown solutions "tropical and temperate forest restoration," "agroforestry," "tree plantations (on degraded land)," "peatland protection and rewetting," and "abandoned farmland restoration" all fit in this land sink category.[34] The carbon sequestration potential of such solutions varies depending on local soil, water, climate, vegetation, and socioeconomic conditions and unfortunately is expected to decline as regional temperatures rise. One study estimated that natural climate solutions could mitigate 21 percent of the greenhouse emissions in the United States.[35]

Although climate scientists used to ignore peatlands—those soggy landscapes where walking can mean sinking up to your knees in wet grasses or sphagnum moss—we now know they store 30 percent of all land-based carbon. Because peatland soils are waterlogged, oxygen-loving bacteria that break down dead plants, and in the process release carbon into the atmosphere, are absent. Instead, the carbon builds up in the peatlands sometimes over thousands of years. When peatlands are drained for agriculture, much of the stored carbon is released to the atmosphere; in fact, drained peatlands account for nearly 10 percent of agricultural emissions. Recent peatland fires in Indonesia and Russia have further depleted these carbon stores as well as raised awareness about the importance of peatlands.[36] The Nature Conservancy, along with government agencies and other nonprofits, has launched efforts to rewet peatlands in the Great Dismal Swamp of Virginia and North Carolina,[37] and the government of Indonesia, spurred by catastrophic fires that left people choking on smoke for months, has attempted to prevent further conversion of peatlands to palm oil plantations.[38]

Reforestation and other land sink solutions can impinge on local livelihoods—for example, when big companies expropriate land from local farmers to establish timber plantations. However, when done right, such solutions can alleviate climate injustice by providing jobs, strengthening civil society, and improving human and environmental health. The nonprofit Coalfield Development retrains unemployed coal workers to install solar and fruit farms on land decimated by mountaintop-removal mining. Asked during a BBC interview whether the retrained coal miners see themselves as part of the Green Revolution, Coalfield Development's founder Brandon Dennison, himself a native West Virginian, responded:

> No, not yet. . . . I think we've not had a clear strategy in coal country for how to seize this transition. . . . From the top, we've not had a lot of support but from bottom up there's a lot of young entrepreneurs just like myself who love this place. We love the beauty of the landscape. We love the vibrant culture. We're committed to staying here and helping our place realize its full potential. And so from the bottom up I'm encouraged. Now we just need some support from the top and when those come together, I think there's actually some real opportunity for coal country.[39]

While known for its coal mining culture and resistance to climate regulations, West Virginia, due to its many creeks, steep slopes, and mountaintop removal of soil and vegetation that can absorb heavy rains, is more vulnerable to the devastating downpours caused by climate change than any other state. Over the next thirty years, there is a better than one in four chance that a third of West Virginia properties will be severely impacted by floods.[40] Just this year, residents watched in horror as raw sewage poured into their homes and contaminated their drinking water supply during floods the likes of which they had never witnessed.[41] Land sink solutions implemented with justice in mind can reduce such flood devastation, while providing sites for growing food, for wildlife, and for sequestering carbon.[42]

Public Goods

> **Maintaining the Earth's climate within habitable boundaries is probably the greatest "public goods game" played by humans.**
>
> —Manfred Milinski, biologist, *Stabilizing the Earth's Climate Is Not a Losing Game*

Under which conditions do people give away their own money to social, environmental, and climate organizations? Research on social dilemmas, public goods, and game theory helps answer this question as do studies that demonstrate how donating money can generate feelings of happiness in donors.

In a social dilemma, individual interests conflict with societal interests—what individuals perceive is best for them is not what's best for society or for the planet. Driving my car is in my short-term interest if I want to go somewhere quickly,

"I can't get enough of these sunny days on the lake."

but if everyone drives their car, we have intolerable traffic jams and rampant greenhouse gas emissions. Donating to a cause can also be conceived as a social dilemma. I would prefer to keep my money and spend it on myself and my family, but if I donate a portion of my income to an organization that supports women's health or peatland restoration, society and our planet (and even my children in the long run) benefit.

Social dilemmas commonly form around public goods such as community forests and a stable atmosphere. A public good is a resource we all share, and yet it is difficult to exclude anyone from taking or using more than their fair share. Unfortunately, rather than cooperating with others to conserve public goods, individuals often select actions that seem to maximize their own short-term benefits, such as driving large cars or eating a lot of beef. Even worse, the so-called polluter elite not only consume more resources through their luxury purchases but also invest vast sums of money in fossil fuel companies and lobby government to lower taxes on their investments.[43] Whether we consider only the affluent or all us as the problem, the joint outcome is bad for everyone: people suffer when they do not cooperate to ensure public goods are shared equitably and conserved for future generations.[44] Humanity exhausts the stable climate that supports all humans and other life on Earth.

Climate change may be the greatest social dilemma or public-goods challenge facing the planet. With over seven billion "actors," the ways in which we avoid

"The Tragedy of 'The Tragedy of the Commons'"

In 1968 the University of California at Santa Barbara professor Garrett Hardin published "The Tragedy of the Commons," a highly cited and influential paper that to this day permeates conservation thinking. Hardin made the case that pastures, forests, and oceans are commons—resources held in common. But because each shepherd and his family benefit from grazing more sheep and each fisherman benefits from catching more fish, overgrazing and overfishing are inevitable, and thus we destroy the commons.[45] Because Hardin reasoned more people means more resource depletion, he further argued that "the freedom [of humans] to breed is intolerable." He went so far as to proclaim that certain races and religions use overbreeding as a means to ensure their own aggrandizement.[46]

Fifty years later, another UC Santa Barbara professor, Matto Mildenberger, wrote a bitter rebuke of his predecessor's claims in a short essay called "The Tragedy of 'The Tragedy of the Commons.'"[47] Mildenberger offers three critiques of Hardin's work, two or which refer to their own tragedies. The third offers hope.

First, while Hardin blames local people for not conserving their resources, the truth is that many indigenous and other communities lived sustainably for years until a fossil fuel or diamond or chocolate company took over their land and converted it to a swath of oil wells, a mine, or a palm oil plantation. Mildenberger claims the same thing happened with our climate: we lived in harmony with our atmosphere for millennia until fossil fuel companies made the decision to ignore the science and push an oil and gas economy. The real tragedy is the destruction wreaked by the fossil fuel industry, mining, and industrial agriculture.

Second, if like Hardin you blame local people for destroying the commons, then the "logical" conclusion is to limit their population or even allow them to perish once they become desperate after being kicked off their land. In a commentary called "Living on a Lifeboat," Hardin uses the metaphor of a lifeboat full of poor people, proclaiming that we should let the passengers drown rather than take them in and allow them to reproduce.[48] He ignores the fact that the rich who have exploited these poor people's resources have impacts on the environment magnitudes greater than the former land occupants now reduced to fleeing their land.

Finally, and on a more hopeful note, Mildenberger refers to the work of the Nobel Prize–winning economist Elinor Ostrom, who spent a career researching the conditions under which communities do manage resources sustainably. In communities that effectively manage a commons, people engage in frequent communication and have tight social networks, which enable them to establish trust and shared norms about resource use. These communities are also able to monitor how resources are being used and have the power to impose rules on resource exploiters.[49]

Not only was Hardin's science wrong in ignoring the many situations where resources held in common did not end up in tragedy, but his science was also influenced by his white nationalist views on immigration and population control. Although his work is still widely cited and used in classrooms today, Hardin is labeled a white nationalist by the Southern Poverty Law Center.[50] We need to explore Ostrom's and others' alternatives to his work that are scientifically sound and offer strategies for conserving our climate in a just manner.

social dilemmas—building trust and shared norms, taking collective action, or even altruism—may not apply. The scale is simply too large. Tree planting, for example, occurs more commonly in smaller communities where people know and trust each other.[51] And the research on what conditions foster cooperation to preserve public goods has usually been conducted in discrete communities or in small experimental groups.[52] In small communities, the public good can be better defined, such as a town forest or public park, and the outcomes of cooperation are more immediate and visible. The townspeople agree to limit cutting so as to conserve the forest for all to take peaceful walks or collect mushrooms and berries.

In contrast, our climate, and the community that needs to conserve it, are planetary in scale. Where in a typical public-goods problem we are mulling over the trade-offs between personal restraint and community benefit today, in our climate public-goods problem we also are considering the impact of our behaviors on future generations whom we will never meet. We are being asked to "cooperate with the future."[53] Compounding the climate issue even further is that it's nearly impossible to calculate the value of the public good—that is, the dollar value of maintaining a climate that supports life on Earth.[54]

Furthermore, the climate system is unfathomably complex. We don't know, for example, if recent heat waves and floods in Seattle and Vancouver or fires in New Mexico are evidence that we have reached a tipping point or simply another climate-related disaster. And it's impossible to measure the impacts of any one person's actions on the future climate, let alone on a wildfire or heat wave that occurs today. What's more, the very notion of responsibility is muddled by lack of agreement on who is causing the crisis. But even if we could somehow agree on the culprits, how useful would that be in taking action?[55]

Not surprisingly, given all this uncertainty, it has been hard to find people who are willing to voluntarily pay for climate solutions. Although many people around the world simply can't afford to pay for climate mitigation, those of us who could afford to pay a climate tax or contribute to an NGO that supports carbon-friendly agriculture often assume that spending money on ourselves—buying nice clothes, going out for a steak dinner, or booking a tropical vacation—will make us happier than contributing to the climate cause. We also may adopt the attitude "Why should I pay money to conserve soils that sequester carbon, when so many of my neighbors aren't contributing a penny?"

And yet people do donate to causes they care about. In 2019 total charitable giving in the United States reached $449.64 billion. Individual giving was by far the largest fraction (69 percent), compared to 17 percent by foundations, 10 percent by bequest (as specified in the will of a deceased person), and 5 percent by corporations. Of the total amount, 3 percent, or $14.16 billion, went to environmental and animal rights groups, whereas 14 percent was spent on education and 9 percent on health.[56] In 2020 climate philanthropy grew nearly five times faster than overall philanthropic giving, aided in part by megadonations such as the $10 billion Bezos Earth Fund. And in 2021 the Donors of Color network launched the Climate Funders Justice Pledge.[57] How can we explain this giving behavior or the conditions that lead people to donate to a cause like climate?

Why and When Do People Donate Money?

> **Bounded self-interest captures the other face of Adam Smith—that people can be selfless.**
>
> —Jason Shogren and Laura Taylor, environmental economists, *On Behavioral-Environmental Economics*

Homo economicus is the idea in philosophy that humans are self-interested and that rational behavior is dictated by a desire for wealth. But how does *homo economicus* explain the finding that in 40 to 60 percent of public-goods experiments, people do not selfishly think only about themselves but also contribute to the public good?[58] Or how do we explain the fact that individuals in the United States donated nearly $310 billion to charities in 2020?[59]

A clue to explaining philanthropy comes from the eighteenth-century economist Adam Smith. Although later anointed as the father of modern capitalism, Smith was not just an economist but also a philosopher. In his *Theory of Moral Sentiments*, Smith wrote that our behaviors are not solely determined by a desire for wealth but also by our capacity for reflection, our sense of justice and compassion, and our need for social approval.[60] Fast-forward three hundred years, and social scientists began using controlled experiments to test Smith's propositions.

If you were once a college student, you may have been asked to enter a windowless room in a university psychology building, where you were given a sum of money and some options, such as keeping the money or donating it to a charity. If this happened to you, you were probably a subject in a public-goods game experiment. Public-goods games shed light on the situations in which people cooperate or, in the case of philanthropy, under what conditions people give away money for a cause. Does being able to discuss options with other players, as opposed to donating in a vacuum, affect giving? Does making players' decisions public affect giving?

In answering these and other questions, public-goods experiments have shown time and again that players do not always act according to their own narrow self-interest. They have confirmed what Adam Smith wrote over two hundred fifty years ago: our self-interest, defined narrowly in economic terms, is bounded by multiple factors in real life.[61] For instance, people make decisions about how much and where to spend or give money based on what their family and friends are doing, loyalty to others who support a cause, or simply who asks them to donate. Decisions are also based on concerns about one's reputation and about fairness, whether people trust nonprofits, and even how giving makes them feel.[62] Not surprisingly, different cultures have different norms that influence philanthropic behaviors.[63] This is all to say that we are not just narrowly selfish and rational beings when we make decisions but also social beings.

We can think of public-goods games as creating temporary social networks with assigned rules that dictate when and how much communication is allowed. The varying communications allowed among players simulate strong and weak ties between people in real-life social networks.[64] When little communication is allowed, the group of players simulates a weak-tie network. Experiments that are repeated with the same players and that allow players to discuss their views more closely resemble real-life networks with ongoing communication and stronger ties.

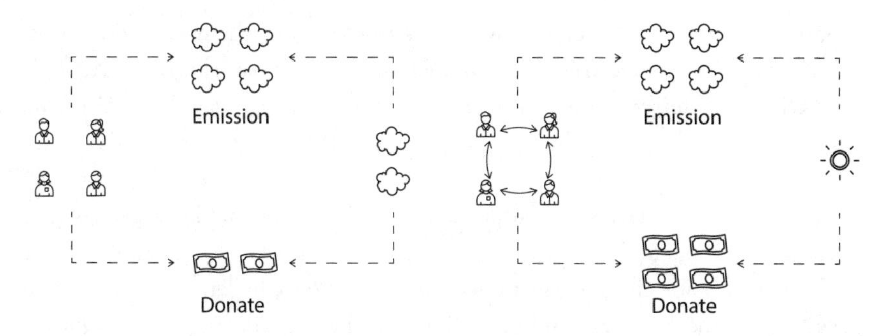

People donate more to a public good when they are able to communicate with others about their donations. Diagram by Xiaoyi Zhu.

Just as players in public-goods games influence each other's giving, people in real life can influence each other's donations to climate justice organizations. A network climate action focused on donations entails influencing friends and family to give money to a climate cause alongside you. But fundraising campaigns often go one step further, asking friends and family not only to donate themselves but also to pass the fundraising campaign message on to their own social networks. Lessons gleaned from research on public goods and from observing fundraising campaigns reveal five rules of thumb that can be used to spread donating behaviors through social networks. Underlying most of these rules is the finding that when people are able to communicate about their giving, they are more likely to give more.

People Give When They See Others Giving

> **Thank you for your donation to Gulf Coast Center For Law & Policy! . . . To make an even bigger impact, spread the word! Tell your friends and family why you support Gulf Coast Center For Law & Policy by sending an email or sharing your support on social media.**
>
> —Executive director, Gulf Coast Center for Law & Policy, email

You may remember the Ice Bucket Challenge—hundreds of videos of people, young and old, jocks and nerds, the gal next door and world-famous celebrities—all grimacing and gasping while pouring buckets of ice water on their heads. Everyone was taking the same action, yet each person was able to customize the action to their own setting and friends and to share their version of the challenge through photos posted on social media. The challenge thus connected participants to friends and to a wider network of people across the United States and around the world. Not everyone who shared a video of themselves doused in ice water or who invited friends to do the same actually contributed to research on ALS (also referred to as Lou Gehrig's disease), which was the goal of the Ice Bucket Challenge. But enough people did that the campaign raised nearly $42 million in three weeks.[65]

In 2012 GivingTuesday was launched as a simple idea: establishing a day that encourages people to do good. Similar to the Ice Bucket Challenge, the goal was to raise money for charity. But unlike the Ice Bucket Challenge, where people gave to the same organization but at different times, GivingTuesday was meant to spur donations to any organization on the Tuesday after Thanksgiving. It was meant as a counterforce to the unbridled post-Thanksgiving consumerism of Black Friday and Cyber Monday. The plan was that any organization could redesign the original GivingTuesday heart logo to fit their particular theme—be it a gay rights rainbow heart or a pink heart associated with breast cancer research. Today GivingTuesday inspires hundreds of millions of people to give, collaborate, and celebrate generosity.[66]

Drawing from their experiences with the Ice Bucket Challenge and GivingTuesday, *New Power* authors Jeremy Heimans and Henry Timms outlined three design principles that are key to making donating or other actions spread via social media campaigns.

- *Actionable*: A fundraising campaign should ask friends and followers to do two things: share the campaign with their friends, family, and followers and donate money.
- *Connected*: A fundraising campaign should promote a feeling of taking action with people you care about and who share your values. It should make participants feel part of a larger like-minded community.
- *Extensible*: The campaign symbols and actions can be readily reshaped or customized by participants. Each person can add their own photos and words of encouragement. Campaigns should be able to incorporate multiple events and communications over time.

Admittedly these "ACE" principles are derived from close observation of campaigns that have gone viral using social media. For most of us, just convincing a few friends to donate to a climate justice organization is an accomplishment. But these principles can push our thinking about donation-focused network climate actions. When asking a friend to donate to a climate justice organization, it's not that hard to ask them to share information about the organization with their friends. Campaigns can leverage the power of tight networks in persuading friends and family to contribute money. They can also exploit the power of social media and messaging apps to communicate about giving with family and friends.

In a sense, social media and messaging are simply expanding on more traditional ways that nonprofit organizations have made donations visible in the hopes of influencing more people to donate—for example, by listing donors' names in their newsletters.[67] Every month my husband used to receive a newsletter from the Finger Lakes Land Trust, which listed categories of contributions (for instance, $500–$1000) followed by a list of those who had contributed at that level. This, however, didn't motivate my husband to jump into those higher donor categories; he continued his same annual contribution. In a more recent newsletter, the land trust simply posted a list of all donors but without the dollar amounts contributed. Perhaps the organization had become aware of a public-goods experiment in which making public the names of all donors, regardless of contribution size, led to an increase in contributions, whereas publishing only the names of donors who had contributed high amounts had no influence on contribution size. The experimenters attributed these results to the prestige of giving large amounts being a less powerful motivator for donating than the shame of not giving at all.[68]

Discussing donations with friends may seem taboo—perhaps even viewed as bragging or virtue signaling to display one's impeccable moral character. Nonetheless, I have found that it's not hard to discuss donations, and nudge people to give, in small groups of family members and friends. During the 2020 US elections, I reached out to friends about which organizations they thought would be most important to contribute to, and my family and I had an ongoing discussion on the Signal messaging app about which senatorial candidates were most likely to benefit from our donations. These informal and ongoing interactions seemed to increase my children's and my giving, much as social interactions among a family network might influence what family members eat. When he was first running for the US Senate, Raphael Warnock leveraged ides about social interactions by asking supporters to set up their own Warnock fundraising site on ActBlue. I called my site "marianneforwarnock." Although the app allowed me to see how much each person was contributing, I promised anyone who contributed that I would not look at this information, as I felt doing so might hamper building trusting relationships. Warnock's campaign demonstrates how social media can serve as a tool for taking action among close networks, making it easier to donate. But the primary mechanism for influence is still conversation with close family and friends.

An article titled "How Content Is Really Shared: Close Friends, Not 'Influencers'" recognizes the power of small, tight-knit groups: "Online sharing, even at viral scale, takes place through many small groups, not via the single status post or tweet of a few influencers. While influential people may be able to reach a wide audience, their impact is short-lived. Content goes viral when it spreads beyond a particular sphere of influence and spreads across the social web via . . . people sharing with their friends."[69] The authors of *New Power* agree: "Today, the most resonant ideas are not those that get flashed at the highest number of people but those that become individualized expressions of affiliation and identity among peers. . . . The good news here? You probably have much more influence over your friends than Kim Kardashian does."[70]

People Care about Their Reputation

Let's go back into that windowless room in the university psychology building where the public-goods experimenter has put you in front of a computer screen. Even though you can't see the other players, you are not playing alone. The computer has just informed you that you are being given twenty dollars. You are asked to allocate however much you want of those twenty dollars to a climate organization (or public good), the remainder of which (if you don't choose to give it all away) you may keep for yourself. How do you make your decision?

"How much are you planning to tip?"

Maybe the first thing you want to know is whether the other players will find out how much you decide to keep for yourself and how much you contribute to the public good. You are concerned about your *social reputation*—you don't like to think of yourself as being selfish. [71] Consequently you decide to give a sizeable amount to the public pool of money. In general, people prefer a self-image that

Free Riders versus Catalytic Cooperation

The climate reporter Robinson Meyer has taken the ideas of social reputation and cooperation to the global scale. He contrasts two approaches to global climate negotiations. Former president Donald Trump's free-rider approach asks why the United States should do anything on climate while China, now the world's largest emitter, continues to burn coal and emit carbon. Free riding happens when groups do not share norms or trust and are not able to monitor each other's behavior. President Joe Biden, in contrast, is taking a coalitional approach, which suggests that climate is not a free-rider problem but rather one of convincing a sufficient number of interest groups that something needs to be done. The current US climate policy discussions are framed in terms of jobs and health to draw multiple interest groups into the coalition.

Meyer goes on to say that China is concerned about its international prestige and will want to show it is attacking the global emissions problem just like the United States and Europe. The coalitional approach would argue that if the United States or the European Union can forge a coalition of interest groups to support climate legislation, it will set norms that other rich and middle-income nations concerned about their social reputation will follow.[72] This might even lead to "catalytic cooperation" among countries. In this scenario, initial accords such as the Paris Agreement spur action in a limited number of countries, in part because they see additional benefits of climate mitigation, including human health. These countries in turn create conditions (such as lower cost of solar as it becomes mass produced or greater knowledge about effective energy policies) that make it easier for subsequent countries to cooperate on climate-mitigation policy in a sort of catalytic chain reaction.[73]

reflects social responsibility rather than self-interestedness, and giving more to a public good can help preserve a positive self-image.

But—beware. In some cases people will actually modify their beliefs to fit their behavior, instead of change their behavior to fit their beliefs. So, for example, you might think you are a good person, but you don't give to climate causes. The resulting "cognitive dissonance" between how you think of yourself and how you act is discomfiting.[74] To resolve the unease, you might decide climate change is a hoax or that your modest gift cannot possibly make a difference—thoughts you use to justify your decision not to give to the climate cause.[75]

Regardless, nonprofit organizations leverage people's concern about their social reputation through extravagant gala fundraisers, at which rich people display their wealth and generosity and where everyone can see who else is there and what they are wearing. Although the year I attended the Finger Lakes Land Trust annual donors' party, attendees were decked out in flowered skirts and Carhartts rather than in diamonds and Rolex watches, and we nibbled not on caviar but on cauliflower, crackers, and creamy dip, we could still envision our social reputation being burnished through attending the event. Social reputation may also play a role in convincing neighbors to join a fundraising campaign. Perhaps concern about their reputation was one reason my neighbors attended an event that I once held at my home to raise money for a climate-friendly congressional candidate.

People Give When They Get Something Back

Telling donors that their contribution will be matched is one way to make people feel they are getting something back from donating. Interestingly, a 1:1 match may be even more effective than a larger match in increasing the likelihood and amount of giving for active donors.[76] In a study of actual donations to the Sierra Club, a promise that a matching gift depended on how much the donor contributed was not effective in raising donor contributions. However, when the match did not depend on the amount contributed, the likelihood of someone giving, and the average amount they gave, rose.[77]

Sometimes donors are given a small gift, which can be effective when the organization's motivations are perceived as authentic rather than manipulative.[78] I periodically find in my mailbox solicitations from the Sierra Club, where a nickel is visibly affixed behind a transparent circle on the envelope. Here the idea is that if Sierra Club gives me a gift, I will be more generous in my donation. This ploy backfires on me because I find it annoying and manipulative. Even though I like the Sierra Club's work, their five-cent gift does not feel authentic.

Yet there are ways to reward people donating to a cause in a more personalized manner. Rewards, even as small as a personalized thank-you, can reinforce the experience of giving as a two-way exchange.[79] Each year I make a small contribution to Rocking the Boat, a youth organization in the Bronx where several of my PhD students have conducted research. I always get back a personalized email

"They gave me this for donating to a solar advocacy nonprofit three years in a row."

from the director thanking me for my donation and expressing his appreciation for the work of my most recent graduate student. This personalized recognition contributes to my feeling of having a meaningful long-term relationship with Rocking the Boat and its director and is more likely than a generic email to make me want to contribute again.

One of my undergraduate students, Margot Schwartz (not real name), used rewards in a fundraiser for the women's health organization Femme International. Schwartz was aware that when members of her sorority spread the word about or contribute to an approved philanthropy, they receive points, which enable then to attend their big end-of-semester formal dance. She coordinated with her sorority's leadership to get Femme International on the list of approved nonprofits. To incentivize participation among the sorority sisters, Schwartz held her two-week fundraiser near the deadline for earning participation points. She ended up raising eight hundred dollars for Femme International. The fact that this was much larger than the amount raised by other students who chose women's health organizations as their network climate action suggests that Schwartz was successful in using rewards that catered to the interests of her sorority sisters.

People Care about Fairness, Transparency, and Trust

Perhaps because it seems fairer when the burden of donating is shared, players in public-goods game experiments give higher amounts when they believe that others contribute.[80] Studies of actual donations to a range of causes, including a public radio station,[81] a national park,[82] and a project to reduce fossil fuel consumption,[83] similarly showed that when people believe that others are contributing to a public good, they are more likely to contribute themselves. This may be due to notions of fairness or wanting to conform to social norms.[84] It may also be due to *conditional cooperation*, or the fact people are willing to cooperate but only if others do too.[85] While it might seem reasonable to worry that people will hold back from making donations if they see that others are already taking care of things by contributing, this is not generally the case.[86]

Being able to openly and cooperatively agree on, and enforce through sanctions, each member's contributions can also promote feelings of trust. In studies of farming communities, when members can determine and monitor each individual's allocation (e.g., of irrigation water) and even punish those who don't live up to their original commitment, they feel they can trust others and are more likely to stick within their allocated amount.[87] Similarly, game experiments have shown that players follow up on their commitment when they jointly come to an agreement on what everyone should contribute and their agreements are binding.[88]

The Intergenerational Goods Game and Climate Futures

The "intergenerational goods game" simulates how climate impacts will be more dire for future generations. Each player is allowed to extract a certain amount from the joint pool of one hundred experimental units. The rules stipulate that if what is left in the joint pool exceeds a threshold amount (players limit how much they take), the original one hundred units are passed on to the next generation of players. This is meant to simulate how, if we limit our climate emissions, we will pass on a healthier climate to our offspring. But if players extract a lot from the joint pool and what is left is lower than the threshold, nothing is passed on to the next generation. In other words, if people are selfish, they pass on to the next generation a climate catastrophe.

So, what were the results of the intergenerational goods game? When each player made their own decision in this experiment, several players extracted excessive amounts, which led to destruction of their joint capital (and the climate!). When, however, decisions were made by a binding vote, the simulated stable climate was sustained and passed on to the next generation of players. The experimenters attributed this result to the fact that the players who cooperated were able to restrain the smaller number of players who might have taken a lot for themselves. Less trusting players felt reassured that they weren't being taken advantage of. However, if the decisions were nonbinding—that is, players agreed on a contribution by voting but then were allowed to decide what they actually contributed without fear of sanctions—more players acted selfishly. This result has implications for policymakers, who can design rules to enable and enforce collective agreements.[89]

People Give Because It Makes Them Feel Good

Research has documented that choosing something to buy can alleviate sadness, a phenomenon known as "retail therapy." Sadness is often a result of feeling unable to control our situation, and being able to choose what to purchase instills a sense of control.[90] But that doesn't mean that buying is the only path to happiness. Perhaps humans are simply wrong when they believe that purchasing a bigger TV or a more luxurious vacation is the only way to make them feel good. Could altruistic behaviors, such as giving money to a charity or environmental nonprofit, also instill feelings of joy and satisfaction and even stimulate the "happiness" portion of our brain?[91] The answer is yes.

We are not surprised when people with biospheric values—those who care about plants and animals—contribute to environmental philanthropies.[92] But people with altruistic values also engage in environmental activism.[93] In my own experience, knowing that their contributions would help poor people weatherize their homes—reflecting an altruistic value—seemed to motivate some colleagues to contribute to the Finger Lakes Climate Fund more than the idea of offsetting their air travel to save the climate (biospheric values). This suggests that one way to influence people to give to environmental causes is to emphasize the human aspects that so many people care about. This is not hard, given the profound suffering, and even deaths, being caused by the climate crisis.

What if altruism is not purely selfless but also motivated by emotional pleasure seeking? In fact, researchers have documented that people who donate to causes, including those focused on the environment, feel personal joy and satisfaction, or a "warm glow."[94] In one study, researchers approached strangers and gave them a five-dollar or twenty-dollar bill to spend by the end of the day—either on themselves or on someone else. The amount of money they received had no bearing on their happiness at the end of the day. Perhaps more surprising, those who spent money on others reported feeling happier. A study of employees who received bonuses similarly found that those who engaged in "prosocial spending"—that is, they bought a gift for someone else or donated money to charity—were happier than those who used money for themselves. This is despite the fact that most people think they will be happier if they spend money on themselves.[95] Further, residents of 120 countries reported more joy if they gave money to others. For example, study participants in South Africa and Canada felt happier when they purchased a goody bag full of treats for a sick child rather than for themselves. The researchers even went so far as to suggest that "the capacity to derive joy from giving might be a universal feature of human psychology."[96]

It seems that the weak correlation of income with happiness can be attributed to the fact that we become habituated to our standard of living, and thus money doesn't play a major role in our day-to-day happiness. To promote lasting happiness in ourselves, what's required is active engagement in meaningful activities. The authors of altruism studies conclude, "Given that people appear to overlook the benefits of prosocial spending, policy interventions that promote prosocial spending—encouraging people to invest income in others rather than in themselves—may be worthwhile in the service of translating increased national wealth into increased national happiness."[97]

A Warm Glow of Giving in Our Brain

The economist James Andreoni's warm-glow giving theory is supported by a study that used functional magnetic resonance imaging to investigate the brain mechanisms behind generous behaviors.[98] Study participants who promised to spend money on others reported feeling happier than those who planned to spend money on themselves. The images showed activity in the striatum, the brain region that rewards behaviors through inducing positive emotions, during the period participants were making generous decisions.[99] In another study of altruism and the brain, people who expressed gratitude—for instance, through "gratitude journaling" or public expressions during holidays or at church—were more likely to donate money to a charity. The scientists connected the giving behavior with activity in the "neural pure altruism" region of the brain.[100]

Emotions can also play a role in online fundraising campaigns. A couple who had worked in Facebook's trust and safety division routinely ran WTF (Want to Fund) Wednesday fundraisers with fellow parents, each time raising about five hundred dollars. Outraged by Trump's family-separation border policy, they decided to devote one week's WTF Wednesday to raising money for a family whose children had been taken away by immigration officials. The WTF Wednesday parents themselves had large groups of friends who worked in trust and safety at tech firms and held altruistic values. As outraged tech workers shared the fundraiser with friends, strangers stepped in to offer 1:1 matches for any donations. Originally hoping to raise fifteen hundred dollars to pay the bond for one immigrant parent, the Reunite an Immigrant Parent with Their Child campaign had reached

$19 million within a week, the vast majority of the money coming from donations under forty dollars.[101]

The emotion unleashed by images of children being separated from their parents at the border may help explain why this campaign went viral. Emotions generated by a movie showing women having to drop out of school because they are too poor to buy menstrual supplies similarly spurred me to donate to my students' Femme International fundraisers. But let's not forget the positive emotions generated by acting with a group of friends on shared altruistic and biocentric values. Although not all of our fundraisers go viral, we can still leverage research about happiness and emotional contagion in influencing friends and family to donate and to share our small donation campaigns.

COVID-19, Catastrophe, and Giving

During the COVID-19 pandemic, a Los Angeles millennial named Caroline Chang found new forms of happiness that did not involve spending money—such as going for walks and playing board games with friends. Although she didn't feel comfortable joining a Black Lives Matter protest because of COVID-19, she began donating twenty-five dollars a month to a nonprofit organization focused on ending racism and police violence.[102] It often takes a crisis to unleash giving behaviors.

In 2021 Americans experienced unprecedented killer heat waves in Oregon and Washington state, never-before-seen killer tornadoes in Kentucky and Tennessee, disastrous floods again in Washington, and raging infernos in California and Colorado. The climate crisis has become a major source of suffering and death in the richest nation in the world. Perhaps not surprisingly, nearly half (45 percent) of Americans reported personally experiencing the effects of global warming, and a majority were worried about harm from extreme weather events in their local area. Slightly fewer (44 percent) reported that their family and friends make at least a moderate amount of effort to reduce global warming.[103]

The year 2020 saw a significant increase in global philanthropic giving to climate change mitigation. Although the amount was still less than 2 percent of total philanthropic giving for the year, the $6 billion to $10 billion donated exceeded pledges made earlier, and more of the funds were directed toward climate justice groups.[104] We can expect that as more and more people suffer from climate disasters, more and more will consider giving to organizations that help alleviate the pain and encouraging their family and friends to give alongside them.

• • •

GIVE: Don'ts and Dos

1. Do explore social media, apps, and digital fundraising tools to engage your family and friends in donating to women's health and education organizations and programs that support carbon sequestration in farms, forests, and peatlands.
2. Do ask your friends and family not just to donate but to spread the word about donating to their friends and extended family. Don't feel shy to ask—they will likely feel happy giving.
3. Do offer recognition to those who donate—perhaps a short personalized note.
4. Don't assume a one-size-fits-all campaign. Create opportunities for your friends to express themselves, such as posting selfies while walking in nature to raise money for a tree-planting and tree-care NGO.

4

VOLUNTEER

There are ways for individuals to plug into system change, and we need individuals to lead systems change and inspire other people to do the same.

—Amy Westervelt, journalist, *Climate One* podcast

Once I learned how much meat consumption and food waste contributed to greenhouse gases, I paid heed to my cooking practices and diet. And influenced by a student in our Climate Solutions course who chose health and education as her network climate action, I donated to Femme International, which supplies much-needed menstrual kits to girls in developing nations so that they can stay in school. But what was I to do about all those other Project Drawdown solutions that do not lend themselves to lifestyle changes or donations? And how might I influence others to get involved?

Many Project Drawdown solutions such as "refrigerant management," "small hydropower," and "utility-scale solar photovoltaics" require government action and international agreements. They also entail changing business, workplace, and NGO priorities and practices. By engaging in the policy process, we *can* support these and other nonlifestyle solutions. And in so doing we turn our attention to systemic or structural change: changing agencies, organizations, laws, and regulations.

From the network-climate-action perspective, influencing policy is not so different from lifestyle actions such as a plant-rich diet—we can take action ourselves, and we can try to influence others to take action. We can support a climate-friendly candidate, and we can try to influence our family to do likewise. We can write letters to our representatives or volunteer for a climate organization and then encourage our friends and family to do the same. And we can share Climate Action Now and other climate advocacy apps with friends.[1] Importantly, we can volunteer for climate-justice nonprofit organizations and invite our friends along. Like other climate actions, policy actions lend themselves to network climate action.

In his book *Politics Is for Power: How to Move beyond Political Hobbyism, Take Action, and Make Real Change*,[2] Eitan Hersh profiles remarkable grassroots leaders who have fomented real change in their community. But on reading the book, I felt discouraged. I was not one of those remarkable leaders profiled by Hersh. In fact, I'm pretty much an unknown in my neighborhood and small city. So, how was I to find a place for myself in the climate movement?

For me, and perhaps for others who are not trailblazers in the climate movement, volunteering for a nonprofit organization whose goal is to influence climate policy makes sense for several reasons. Joining an organization enables you to expand your sphere of influence from lifestyle behaviors to climate policy on anything from wind farms to state regulations that ban food waste from landfills. Whether they are an informal online social media group or a registered nonprofit, groups and organizations inform their members about strategic actions—such as when and to whom to write letters on legislation being considered by government and the most effective way to submit comments on proposed rules and regulations. And by informing and mobilizing members at critical times, organizations can have impacts greater than the sum of the impacts of each member acting alone. What's more, organizations with credibility provide access to the policy process; legislators look to well-informed members of organizations in formulating policy.[3]

For those interested in reducing food waste and encouraging a plant-rich diet, working with a nonprofit organization can help bring about the structural changes that make these lifestyle behaviors easier for large numbers of citizens. One of our Cornell students worked with the Friendship Donations Network, a local food-recovery nonprofit, to devise a system through which his fraternity could donate the brothers' uneaten food. If he were to alter his own fraternity's food-waste practices, he might create a path for other fraternities and sororities across the campus to do the same. Likewise, should volunteers with Elders Climate Action help to get legislation passed that provides a grid of electric car recharging stations, we will have made driving electric vehicles immeasurably easier and may even help US consumers reach a tipping point in their car-buying habits.

The diversity of climate groups—from large established organizations like the Climate Reality Project, to activist groups like the Sunrise Movement, to informal Facebook groups—creates myriad opportunities for volunteers to become involved in activities that reflect their interests and time constraints. Someone might take on the role of volunteer social media coordinator, join a committee urging local government to divest from fossil fuels, or help organize action parties on Zoom. They might find a way to contribute their skills—for example, writing newsletters or doing legal work—or share their knowledge about local government, finance, social justice, or transportation. Importantly, volunteering for an organization not only provides access to the policy process and greater impact than we could have acting alone, it also affords us opportunities to use our skills and knowledge in ways that are meaningful and enjoyable.

Whereas the term "volunteer" is often associated with charity and other non-political activities, in the end much of the work of the climate justice and other social movements depends on the skills, expertise, and energy of volunteers.[4] Nearly anyone can volunteer—even those of us who are not inspirational leaders or celebrity influencers.

My Journey toward Climate Activism

> **The capacity for sustained collective action is conditioned mainly by the presence of established institutions and organizations.**
>
> —Robert Sampson, sociologist, "Civil Society Reconsidered"

For years I focused on two things: family and work. I was aware of climate change but found it too depressing to think about. I tried to convince myself that climate change would simply go away. But once my children grew up and left home and climate change lunged into climate crisis and impending climate catastrophe, I found myself with more time and feeling an increased urgency to do something. I was what political scientists call "biographically available"—at a time in my life when I was less encumbered by family responsibilities and more able to volunteer.[5]

Political Hobbyism

Eitan Hersh calls them "political hobbyists"—those people who constantly fall for clickbait headlines, only to become outraged by the latest political shit show. After Donald Trump was elected in 2016, I instantly became a political hobbyist, fascinated by, even addicted to, the news. It was like a twenty-four-hour soap opera or, perhaps literally, a reality TV show. In his 2020 primer on how to be more politically effective, Hersh urges people like me to stop clicking on outrageous headlines and instead use that time to do something useful to change what's behind the banners.[6]

I still click on the clickbait at least once a day, but I am less likely to fall down a rabbit hole of news story after news story, which has freed up my energy for more meaningful activism. I also repost climate stories to social media, making me an "Internet activist."[7] While some belittle such small actions as "slacktivism," or activities worthless for anything but making you feel good,[8] slacktivism may not be all bad. After all, social media posts can raise awareness about particular issues and policies. By lowering the costs of activism, social media can also be a first step to more substantive participation. It allows people to try out activist identities and different ways to express their concerns and to connect with other activists.[9] But I knew that I wanted to do more than click and share news articles.

Network Climate Action

As I became more and more concerned about the climate, the notion of people influencing their family, friends, and colleagues—their social networks—took center stage in my ruminations. I decided to launch the Cornell Climate Online Fellowship, in which fellows from countries around the world would implement a network climate action—that is, they would choose a Project Drawdown climate action and persuade a close group of family or friends to take that action alongside them. As I compiled research about social influence for the fellows, I realized how difficult taking a network climate action was. I felt like a hypocrite asking fellows from much less privileged backgrounds than my own to do a network climate action when I had never tried to implement one myself.

I decided to launch my own. My action would be buying air travel offsets, and my network would be my faculty colleagues. I work in the Cornell University Department of Natural Resources and the Environment, so I figured it would be pretty easy to convince my "green" colleagues to buy air travel offsets. I also assumed that the ability to contribute to a local organization, the Finger Lakes Climate Fund,[10] would be attractive to my colleagues. The organization's website asks the user to enter their travel itinerary and then calculates the resultant CO_2 emissions and the suggested donation for those emissions. The contributions are used to insulate the homes of low-income county residents. Whereas there are national and even global carbon offsets programs, in which the money is used to support tree planting and other ways of drawing down CO_2 in the atmosphere, I thought supporting a local offsets program might help alleviate my colleagues' fears about the credibility and worthiness of a more distant offsets program.

"Guys, I've signed us up for a climate hunger strike."

I launched my network climate action with a short presentation at a faculty meeting about our local Fingers Lakes Climate Fund air travel offsets program. Then I sent out an email to colleagues with instructions for how to contribute to the fund. Applying my understanding of identity and social norms, I created a "Carbon Race" team—a group of individuals whose collective donations are listed on the Finger Lakes Climate Fund website, along with the team's ranking based on how much each team has contributed. I named our team "Fernow Hall" after our campus office building. My hope was that Fernowites would identify with the team and want to express their Fernow and environmental identities.

The faculty response was at first underwhelming. I pressured my husband and adult children to contribute as part of our team and renamed us "Fernow and Friends" to reflect the growing number of participants who didn't work in Fernow Hall. Next, I sent dynamic norm messages to faculty about how many people were already contributing and how our Fernow and Friends team had moved to sixth place in the Climate Fund's Carbon Race.[11] I tried to engage my colleagues in informal conversations but felt uncomfortable when I knocked on their doors to try to convince them to offset their travel. To give my actions cover and lessen my unease, I shared that I was doing this as a project for a fellowship program that I was directing.

I also constantly reminded my colleagues that the Finger Lakes Climate Fund used our donations to help local low-income families weatherize their homes. This connection with local families seemed to sway some colleagues' interest more than the opportunity to offset their travel for climate reasons. Despite the fact that we are environmental professionals and I had assumed that appealing to environmental or biocentric values would be a winning strategy, it turned out that drawing on altruistic values was also effective.[12]

To my surprise, I discovered that graduate students, whom I hadn't targeted because of their meager salaries, were interested in offsetting their travel. Their interest gave a boost to my effort. I tried to make the offsets activity fun by donating travel offsets to our department's annual auction to raise money for graduate students. The students became allies for my campaign and additional messengers for the travel offsets message.

Applying PANIC Principles to My Network Climate Action

The PANIC principles served as a useful guide in my effort to mobilize colleagues to buy air travel offsets.[13] To these five principles I would add "adaptive"—constantly readjusting one's strategy as one learns, for example, about colleagues' values or that a new ally (e.g., graduate students) can be messengers for the campaign.

***P**ersonal:* Personal interactions are more effective than generic appeals. Think about the influence of those generic flyers you find in your mailbox versus talking to someone in person. I used personalized messages drawing on what I knew about my colleagues' interests and often stopped by their offices to chat about the Finger Lakes Climate Fund and their travel offsets program.

***A**ccountability:* When people can observe each other's behavior, they feel more accountable and want to build a positive reputation. The Finger Lakes Climate Fund website allows you to create a team of people who have offset their carbon emissions, and it lists team members' names on the site.

***N**ormative:* Normative refers to what people believe others think they *should* do (injunctive norm), what people observe others actually *doing* (descriptive norm), and what behaviors they think are becoming *more common* (dynamic norm). I was not able to deploy a descriptive norm, which is used when a majority of people are already doing the targeted behavior. So, I asked family members and friends to join our team, which enabled me to send dynamic norm messages indicating that more and more people were offsetting their travel.

***I**dentity relevant*: We are more likely to mobilize people to take an action that reflects their identities. In my department, people consider themselves to be environmentalists, so I used this identity in asking them to contribute to the climate fund. I also used what is called a "foot in the door" strategy.[14] I tried to get people to join our Fernow and Friends Carbon Race team no matter how little they contributed. In some cases, I even reimbursed them. The idea is that once a person has taken a small step toward offsetting, they will start identifying as an ethical traveler and be more likely to engage in the behavior in the future.

***C**onnected:* As more people joined our Fernow and Friends team, I hoped to be able to leverage our members' existing social networks so that people would hear the offsets message multiple times from multiple colleagues.[15]

Due to a lot of persuading on my part, and my own donations, our Fernow and Friends team had thirty-four members, was in second place in the Carbon Race, and had offset 337,848 pounds of carbon after about six months. (Once the pandemic hit and hardly anyone was flying, I stopped actively recruiting team members or soliciting donations.)

My campaign did not fulfill my expectation that all my environmentally conscious colleagues would readily jump on board, but it did demonstrate that with a lot of hard effort I was able to persuade a good number of colleagues, family members, and graduate students to join a network climate action involving donations. Perhaps in the future I can apply lessons learned from public-goods experiments, such as the importance of social reputation and rewards (see chapter 3). And now that we have a small group of university travel "offsetters," we might expect the university to be more receptive to a university travel offsets policy.[16]

In fact, as a result of my department offsets experiment, I was invited to join Cornell's low-carbon travel working group, whose mission is to draw up policies for reducing the university's travel carbon footprint. It may turn out that the impact of my network climate action was not so much our team's second-place standing in the Carbon Race but rather that my action catapulted me to become part of a team developing university-wide policy. And thanks to COVID-19–induced travel bans, our working group was able to capture lessons learned about how to conduct meaningful workshops, meetings, conferences, and even research, as well as collaborate with civil society organizations,[17] without voyaging across the ocean, the country, or even to the other side of town. It turns out that many people—such as those with physical disabilities, budget constraints, and family responsibilities—prefer remote options, suggesting that there are equity issues to consider in university climate- and people-friendly travel policies.

Discovering Volunteerism

Having tried out the travel offsets network climate action in my professional life, I wanted to become more active as a volunteer in my off hours. I wrote to organizations and offered my services, trying to find a volunteer opportunity where I could use my professional expertise. But nobody was writing me back—perhaps because I was reaching out to expert-driven or professional organizations, whose members simply donate, instead of more grassroots organizations that welcome members as volunteers.[18] Finally, I got a reply—from the nonprofit Elders Climate Action. Unlike many of the bigger environmental organizations, Elders Climate Action is volunteer-driven, although its volunteers have significant expertise from their long professional careers. Elders Climate Action encourages members to conduct education programs and policy actions and thus seemed ideal for people like me who want to contribute their expertise and see the impact of their volunteer efforts on policy.

I talked over the phone with the Elders Climate Action national leader, who had responded to my email. Maybe I could be a speaker for one of their webinars, I suggested, given I had initiated and implemented the Cornell Climate Online Fellowship, taught network climate action courses to global audiences and Cornell students, and coauthored a book on climate communication. But that wasn't happening—and I could sense that self-promotion was not going to work for me in this organization. I became humbler while still seeking a role that would draw on my skills and experience.

When people look for an organization to volunteer with, their choice is often driven by their preferred contribution, such as fund-raising for their church or providing meals for the homeless. They also seek a role that reflects their personal values, identity, and skills and that alllows them to feel worthwhile, connect with others and feel a sense of belonging, or enhance their career and other volunteer prospects. Furthermore, volunteers seek roles that are challenging and

"I can fly and see through walls and use Excel."

fun.[19] Many of us, in fact, may have an image of our ideal role as a volunteer (see chapter 5)[20]. At Elders Climate Action, my ideal volunteer role was to draw on my expertise (rather than do more mundane tasks) while learning from other members about how to effectively advocate for climate legislation. I also wanted to be a long-term volunteer rather than simply help with one event. I knew that to stick with the organization, I needed to feel energized by what I was contributing and learning.

My search for a role in Elders Climate Action began with filling out an application to launch a New York State chapter since New York was glaringly absent from the organization's Council of Chapters. As I went through the form, I realized that I needed a cochair. Because I didn't have any friends who I thought would be willing to do this, I appealed to the Elders Climate Action national cochair for advice. She told me how she'd had to meet repeatedly over the course of months with local elders before being ready to consider launching a local chapter. I wanted to do something now—not after months of meetings. Furthermore, I was aware Elders Climate Action valued in-person gatherings for state and local chapters, but I did not want to organize such meetings. New York is a big state and even if we formed an Upstate New York chapter, in-person meetings would involve a lot of travel, which struck me as morally inconsistent with working to draw down greenhouse gas emissions. I have taught online courses for years as well as facilitated Facebook and WhatsApp learning communities, so I felt comfortable doing things remotely. I eventually realized that forming a local or New York State chapter wouldn't be the ideal role for me. Knowing myself, I could predict that if I became too frustrated or didn't enjoy what I was doing, I would drop out after a while.

Next, I turned to the Elders Climate Action group that supported the Environmental Voter Project through text-banking. This involves using an app to automatically call or text-message voters inclined to support environmental policies but who have not participated in recent elections.[21] I was fascinated by how the Environmental Voter Project uses big data to identify likely environmental voters according to their demographics. Through massive surveys, they have discovered that older Latina women, for example, are likely to be concerned about the environment. Using publicly available voting records of any older Latina woman within a targeted swing state or district, they make phone lists, and volunteers call or send text messages to older Latina women (or to other individuals who are in a demographic that holds positive environmental attitudes but who have a poor history of going to the polls). I tried this once. I downloaded an app and pressed a button that sends fifty messages at once asking people if they plan to vote in an upcoming election. Most people don't respond, but a few send thank-yous or say they are indeed planning to vote. A few people write back nasty responses. This made me feel uncomfortable. Earlier, while helping with a congressional campaign, I learned that I hated knocking on doors to canvas for political candidates and that I hated calling people's phones to urge them to vote. Now I found I even disliked texting people. It felt intrusive, and I couldn't manage to simply brush off the few caustic responses I'd received. Reluctantly I decided this wasn't working for me.

I decided to share information about the Environmental Voter Project with a friend who is a very outgoing retired teacher and who ended up becoming a super caller and texter for the project. The occasional negative replies don't bother her. She enjoys conversations with interested voters. She feels she helps those on the other end of the call or text by just providing someone to talk to or helping them know where and when they can vote. My friend has found a volunteer role that works for her.

I was fortunate that as I was searching for a rewarding role, Elders Climate Action decided to launch an education committee. I volunteered to chair the committee but was told that the selection of a chair needed to be a democratic decision. When the organization's national leader asked during a meeting who wanted to chair the committee, someone else volunteered. Then another. My heart sank, but I kept quiet because it's not in my nature to be assertive. I had the feeling that if I weren't chair, I might get frustrated with the committee's direction and drop out. I would not support, for example, simply giving people information about the climate crisis since research shows that knowledge doesn't necessarily lead to action. Finally, the leader asked me if I would like to chair, and I also said yes. In the end, we decided on three education committee cochairs. At last I was on my way to finding a volunteer role in which I would, hopefully, feel worthwhile—a role based on my expertise and interests and that would allow me to use my years as an educator and education researcher to push others in their thinking about educational strategies that lead to action.

As the climate philosopher Elizabeth Cripps has noted, you may need to revise your volunteering decisions as things change. As time went on, I became more and more interested in food and farm policy and lost interest in the education committee activities. I stopped pulling my weight as cochair. I resigned from the education committee and hopped over to the policy committee, where I am working on an organizational sign-on letter to advocate for climate-friendly policies in the US Farm Bill—a massive bill that will have significant impacts on methane emissions. I also began writing educational "flash quizzes" and food and farm advocacy actions for the Climate Action Now app. Users of the app register with their name and address and choose from thousands of educational and advocacy actions. The advocacy actions are prewritten letters that are automatically addressed to the user's state or federal representatives. I also organize "action parties" where experts present on a climate topic and users take related actions on the app. Not surprisingly, my action parties focus on food and farm policy. At one such party, *Guardian* journalist George Monbiot talked about his new book, *Regenesis*. Participants then sent emails to Congress supporting plant-rich diets, perennial agriculture, and alt-protein research and development.

Deciding Which Volunteer Groups to Join

The climate philosopher Elizabeth Cripps has outlined four steps for deciding how to choose a volunteer role that works for you:

1. Assess existing climate groups and organizations. How might you work with them, and how might they work together?
2. Decide where to devote one's efforts, taking into account your skill set, the cost of the action, and your potential influence as well as that of a group you might choose to work with.
3. Decide where to allocate time based on knowledge about opportunities to join activist groups, including their communication style, needs, and priorities.
4. Be flexible. You may revise decisions as you learn new things about yourself and the groups you are working with.[22]

Harking back to our discussion of practice theory in chapter 2, the app is a "thing" that enables climate-concerned citizens to quickly contact their representatives and thus transforms letter writing from something tedious to something that can be done with a couple of clicks on your phone. This in turn is transforming advocacy practices of organizations using the app, such as Elders Climate Action and the Climate Reality Project, enabling their members to write tens of thousands rather than hundreds of letters to their representatives.

Peripheral and Core Members: Finding Your Place as an Activist

> **It is worth distinguishing between the social networks of ordinary life (relatives, friends, coworkers) and the networks created intentionally by movement activists to further their goals.**
>
> —Kenneth Andrews and Michael Bigg, sociologists, "The Dynamics of Protest Diffusion"

For some, an ideal volunteer role is being super active and becoming a leader on committees and in the overall organization. These people are "core members" or "core actors" at the heart of an organization or social movement.[23].

Core actors are crucial to the success of organizations. They can also form a cadre of activists that determines the success of social movements. During the civil rights movement, sit-ins occurred at drugstore counters in some southern US cities but not in others. What explained the difference between cities? It turned out that the particular organization, such as the NAACP or the Southern Christian Leadership Conference, was not important. Neither was the number of local members of any such organization. Rather, the most important factor predicting whether sit-ins occurred in any one city was the presence of a cadre of activists, what we might call core actors, associated with one of the civil rights organizations.[24]

But what about the majority of volunteers who are not core members, possibly because they have less interest or less time to be active? What about volunteers who might come out to a onetime event such as a protest or a tree-planting day or who occasionally listen in on a webinar or share a post on social media? They are more than slacktivists, but they are not organizational leaders or even core actors. They are "peripheral members" of an organization or movement.[25]

Through their ability to access the expertise of core members and staff, peripheral members can become a powerful force in influencing climate policy. At Elders Climate Action, for example, we have core members who worked for the Environmental Protection Agency, the Department of Defense, and city government and who, in retirement, guide our policy advocacy. A much larger group of peripheral members help populate events such as letter-writing campaigns, where there is power in numbers.[26] I recently joined my local Finger Lakes chapter of the Climate Reality Project, the global climate organization started by Al Gore. As a peripheral member, I do my best to follow their weekly alerts—sent to us by core members—which tell me what to say to which congressional or state representatives and when to call or email. I used to enjoy our Climate Reality tweetstorms, where volunteers posted, liked, and forwarded as many proclimate tweets to congressmembers as they could, while chatting amiably and telling stupid jokes with fellow climate tweeters on Zoom. These one-hour events were not

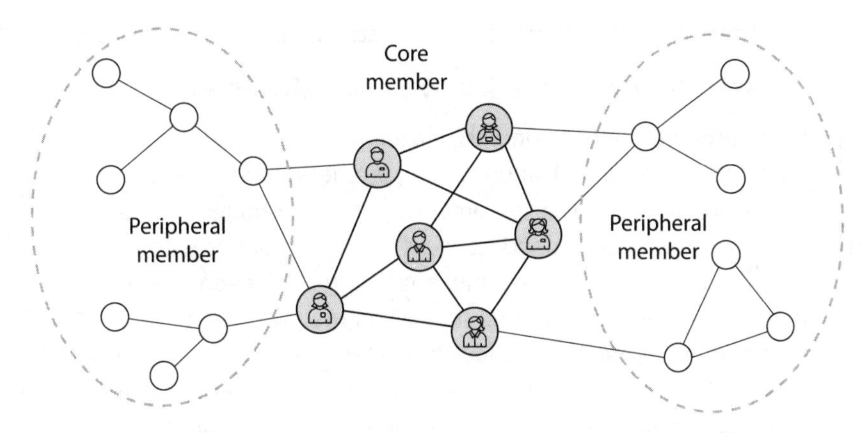

Core and peripheral members play different roles in organizations and social movements. Diagram by Xiaoyi Zhu.

only social and fun. Our first tweetstorm resulted in our climate tweets being seen by over seven million people. (It was estimated that the cost of the publicity we generated would have been $65,000 if we had hired an advertising firm.) Unfortunately, we generated so much traffic that Twitter decided we were bots, and we had to curtail our tweetstorms.

Peripheral members can also provide legitimacy for a young organization, especially if they are simultaneously members of another older and well-regarded organization.[27] Imagine a volunteer with a respected social services nonprofit that provides meals to unhoused people in their community. They decide to join a local climate organization because they see connections between feeding the unhoused and reducing food waste and greenhouse gas emissions. Whereas they may be a peripheral member of the climate organization, the fact that they are associated with a well-respected social services nonprofit can bring credibility to the newly established climate organization.

The activity of lots of peripheral participants kept the climate movement alive during times when many elected officials were climate deniers and blocked climate-friendly legislation. Activities peripheral members engaged in, such as webinars or talking with neighbors, helped build the social and human capital and shared identity that were needed for groups to respond when public attention turned to climate change.[28] Now that society, the media, Congress, and the president are focused on the climate, organizations are calling on this large pool of volunteers to call, email, tweet, and meet with their representatives to let them know there is support for climate legislation being enacted now.

It's true that you don't have to be a core or even peripheral volunteer with an organization to engage in climate action. Lots of people are reducing meat

Organizational Bridgers and Social Movements

Peripheral members of one organization are often core members of another organization. Such peripheral members form a "bridge" between the two organizations. The bridge allows exchange of ideas, expanding the thinking of both groups while allowing them to hone their message so that it resonates with different audiences.[29] During the civil rights protests, pastors of Black churches helped to mobilize parishioners and thus served as a bridge between churches and civil rights movement organizations.[30] Today we see active church volunteers similarly bridging from their church to organizations fighting for our climate.

When several members bridge two different groups—for example, three members of a local mosque are also members of the local Climate Reality Project chapter—they form "thick" bridges between the two organizations. Thick bridges are more likely than "thin," or one-person, bridges to engage members of one group in activities of the other.[31] Multiple ties, or thick bridges, between organizations allow coordination of protests, boycotts, and other social movement activities.[32] In this way, thick bridges can scale up the impact of single organizations and make a more powerful climate justice movement.

Because the backbone of the climate justice movement is a network of organizations whose goals complement each other, such as Elders Climate Action and the People's Justice Council, members who are able to bridge across organizations are critical.[33] The protests in 2016 against the Dakota Access Pipeline on the Standing Rock Reservation in North Dakota provide an inspiring example of how coalitions across different groups can lead to policy changes. The protests brought together American Indian and environmental groups, whose members withstood bitter cold, being blasted with water hoses, and other violence from authorities to show the world the connection between oil and gas pipelines, water, climate, and indigenous rights.[34] The protests helped persuade President Barack Obama to rescind the pipeline's permit. Although President Trump reversed President Obama's order, a subsequent decision in the courts again halted construction. (Participation in the Standing Rock protest also inspired Alexandria Ocasio-Cortez to run for Congress.[35])

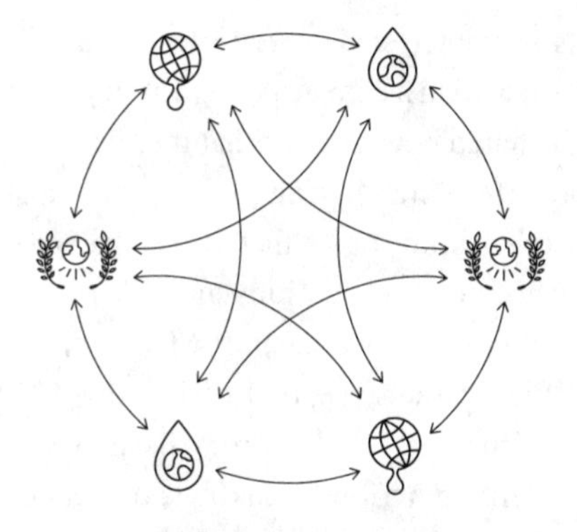

Climate organizations connected to each other

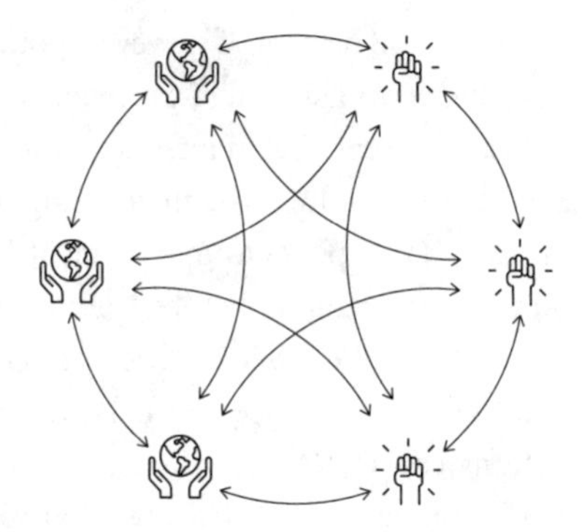

Climate organizations connected with social justice organizations

Climate organizations will be more effective if they coordinate with other climate organizations. Climate organizations will need to connect with social justice organizations to ensure equitable climate solutions. Diagrams by Xiaoyi Zhu.

consumption, for example. In Germany, a country that loves its bratwurst and kebabs, more than 40 percent of the population recently reported cutting down on meat, and the new plant-based "meats"—such as the Impossible Burger—are becoming a staple at fast-food joints.[36] In barbecue-crazed America, nearly a quarter of the population reported eating less meat during 2019.[37] These meat-reducers are not necessarily members of an identified group such as an animal welfare or climate organization. Rather, they might be thought of as a "latent" or loose network of people with similar interests.[38] Even in the absence of coordinated action, as more people turn to plant-rich meals, our restaurants and grocery stores, and soon our food producers and farmers, will meet the consumer demand for meat-free products. In this way, these uncoordinated meat-reducers can and do influence the food industry and the climate. But more formal networks and nonprofit organizations can coordinate the collective efforts and talents of many individuals to move policy change more quickly.

Building Efficacy

> **Onward and upward, and let the efficacy force be with you.**
>
> —Albert Bandura, psychologist

In 2021 Cornell student Shelby Haley conducted a survey and interviewed Elders Climate Action members. She wanted to find out what kind of political activism they were engaged with and what they saw as barriers to becoming more active. Haley found that 98 percent of survey respondents were interested or very interested in climate change policy and 92 percent had participated in a climate policy action within the last two years. Yet in interviews the elders shared how they felt that they had limited ability to influence climate change policy. They found trying to change national climate change policy as an individual was overwhelming or felt that their single voice was not impactful or that they did not know how to effectively influence others.

In short, the elders lacked a sense of efficacy: the belief that what they do can make a difference. Some lacked self-efficacy, the sense that their individual actions can effectively change the climate situation.[39] Others may have lacked participative efficacy. They see that Elders Climate Action is making a difference but doubt that their participation will contribute to the outcome already being achieved.[40] Lacking a sense of efficacy, these elders are likely to ignore the organization's calls to write a letter to Congress or join a divestment campaign.

Albert Bandura was ranked as the fourth most influential psychologist of the twentieth century (behind B. F. Skinner, Sigmund Freud, and Jean Piaget). While completing his education in rural Alberta, Canada, at a high school with a total of two teachers, he realized that whereas "the content of most textbooks is perishable . . . the tools of self-directedness serve one well over time."[41] He spent much of the rest of his life converting this realization into research about self-efficacy.[42]

Self-efficacy impacts what goals we choose to pursue and what actions we take. A person with low self-efficacy or low participative, collective, or political efficacy believes they or the group they belong to cannot accomplish things and thus does not even try. But how does one foster self-efficacy, or, perhaps more important for climate actors, how does one foster political efficacy, the belief in our abilities to understand the political realm and act effectively within it?[43]

Types of Efficacy

Self-efficacy: belief in one's ability to succeed in specific situations or accomplish a task[44]

Collective efficacy: a group's shared belief in its collective abilities to organize and execute the courses of action required to reach goals[45]

Participative efficacy: belief that an individual's contribution is important to the success of collective action[46]

Political efficacy: belief in our abilities to understand the political realm and act effectively in it[47]

Imagine a young person who has nervously ventured out to their first Extinction Rebellion "die-in" protest.[48] They return home that night feeling as if they succeeded in helping to block traffic along an important bridge and have braved the motorists and police angry at them for lying down in the road to obstruct traffic. They mastered their fears and didn't chicken out at the last minute.

This person has had a "mastery experience," an important strategy to enhance efficacy. Mastery experiences allow individuals to master a skill or behavior, such as participating in a nonviolent, disruptive protest. "Vicarious experiences," or seeing people one admires take action and achieve success, can also build efficacy. Perhaps before venturing out to their first protest, the young person was able to observe and interact with other protestors, which helped them feel as if they too could succeed in a nonviolent, obstructive protest. We can imagine the other protesters expressing faith in the novice's abilities, coaching them through any anxieties and offering them a sense of the power that comes from working in a group rather than alone. The experienced protesters help the novice appreciate the value of the "sense of dignity, community, and solidarity that can come from an active political life."[49] Such "social interactions" and "attention to emotions," along with mastery and vicarious experiences, build self and political efficacy.[50]

But how can one maintain a sense of efficacy, let alone hope, when we face new and more devasting climate disasters seemingly every week? Last year, as my daughter drove from my home in Ithaca after the holidays to her home in Denver, she was shaken by yet another unforeseen disaster—this time the fires around Boulder, Colorado, that turned nearly a thousand suburban homes, hotels, and shopping malls into an unprecedented inferno. What could I do to help her deal with her despair?

The highly acclaimed marine biologist and climate communicator Ayana Elizabeth Johnson thinks that being able to avoid environmental despair and keep moving forward depends in part on one's psychological makeup. Those who are able to keep active are better able to survive. She puts it this way:

> I feel very lucky that my brain chemistry doesn't tend towards anxiety and depression. That is just a luck of the draw of how I was constructed at birth. And I know that it's much harder to do this work if you're more susceptible to those things. And so, I'm really grateful that I have a mind that tends towards "It's really bad, what do we gonna do, who's the team, what's the strategy?" "Let me make a checklist, like let me just chip away at this" is my immediate instinct when faced with even a problem of the magnitude of the climate crisis. But even I am stopped in my tracks by the horrors of the news.[51]

Self-Efficacy

People with a strong sense of self-efficacy

- develop a deep interest in the activities in which they participate,
- form a strong sense of commitment to their interests and activities,
- recover quickly from setbacks and disappointments, and
- view challenging problems as tasks to be mastered.

People with a weak sense of self-efficacy

- avoid challenging tasks,
- believe that difficult tasks and situations are beyond their capabilities,
- focus on personal failings and negative outcomes, and
- quickly lose confidence in personal abilities.[52]

What Keeps Us Going? A Tale of Two Volunteers

Whatever our psychological makeup, we need the small successes or so-called mastery experiences—alongside vicarious experiences, social interactions, and positive emotions—to develop a sense of efficacy. We also need these experiences to keep us going as volunteers, as illustrated by the stories of two longtime activists.

From Occupy Wall Street to Reclaim Philadelphia: Small Wins

If you land on the Wikipedia page for Occupy Wall Street, you will see a photo with my former master's degree student Danny Rosenberg Daneri in the center. Daneri's activism started in high school when he did a two-week stint with Earth Day Network, cleaning up trash in Washington, DC. During his first year at Oberlin College in Ohio, he was spurred into activism by the COP meetings in Copenhagen and Bill McKibben's 350.org university divestment campaign. At the time, a coal plant next to the student union in the middle of Oberlin's campus provided heat for college buildings. Daneri helped found the Coal Working Group, which succeeded in forcing the coal plant to shut down. The student group then turned their efforts away from campus protest to the surrounding community. They joined forces with police and local government officials to provide free bike repairs, and even free bikes, helmets, and lights, for residents of Lorraine County.

By 2011, when Adbusters' call for protests against Wall Street spread across the Oberlin campus via Facebook, Daneri had taken his share of political science courses and had campaigned against the Tea Party movement and for workers' rights. He had acquired organizing skills and was beginning to connect the labor, justice, and environmental movements. So, after coordinating with activist groups that agreed to fund their trip, he and other activists climbed into a van bound for Wall Street. Looking back, Daneri says, "The biggest thing that happened at Occupy Wall Street was the networks that were created. There were so many like-minded people. The online forums persisted and propelled a new series of actions. In Ohio, people started connecting over fracking and environmental issues."

Fast forward to 2022. When he's not studying toward his PhD in political science at Princeton University, Daneri volunteers with Reclaim Philadelphia. He is peripherally involved with the group's climate justice work and focuses his core efforts on housing and labor. Daneri feels that unlike climate, housing is an issue where you can have small but tangible wins, such as rent control legislation. Local government has a large say in what happens in housing, and it is easier to influence local elected officials than to influence a massive, uncoordinated global climate governance system (as was all too evident at the COP26 meetings). When working on housing, Daneri also can observe an immediate improvement in people's lives: the rent goes down, people are better off. It's harder to see how working on climate immediately improves people's lives. You can put a solar panel on a roof, but its impact is only realized if millions of other people also install solar panels. There is no immediate improvement in the finances or health of the person who had the panels installed. But if you raise the minimum wage, people experience immediate benefits. Daneri's wanting to make concrete changes in people's lives explains why he is a core member of Philadelphia's housing and labor movement but only peripherally involved with Reclaim Philadelphia's climate activism.

As a core organizer responsible for training other activists, Daneri talks about empowering people to realize a collective vision of a cause through small wins. He also mentions asking people to take on more responsibility: if you are planning to write a press release, don't just gather those with experience. Bring in a new person so they can learn alongside the others—or, in Bandura's words, so they can have a mastery experience and acquire self-efficacy.

When asked whether he feels he is making a difference, Daneri shares, "Whenever I work on any one project, I am not particularly optimistic. I think there is a small chance we will succeed. But it's a chance worth taking. When we worked on fracking, did I ever think we would eliminate fracking? Perhaps there was an outside chance. But we could succeed in shutting down some wells. And in making wells safer, preventing one stream or community from being polluted. We might not win the war, but the battles are worth fighting."

Daneri closed our conversation by reflecting, "People need success to maintain hope. Intermediate wins are important. You can't maintain hope in a campaign that is a series of losses. The wins can be small but there need to be wins. A campaign that isn't scaffolded so that you have opportunities for small wins along the way is not a good campaign. Small wins and my moral compass keep me going." And echoing the words of Greta Thunberg talking about the friends she made through the climate movement,[53] Daneri concludes: "It's also about community. It's nice to find and interact with people who care about the same things I do. Activism has been my primary way of meeting people and making friends for my whole adult life. There's a lot of community that's built when people work on meaningful things together."

Addressing the Climate Crisis from Many Angles

I first met Tom Hirasuna when he took my Network Climate Action online course. Although our students were from over fifty countries around the world, I soon learned that Hirasuna lived only a couple of miles from me and that our kids had been friends growing up. Unlike me, Hirasuna had somehow managed to volunteer while his children were young, serving on the board of our Tompkins County branch of Cornell Cooperative Extension. It wasn't until 2017, after watching *An Inconvenient Sequel: Truth to Power*, that Hirasuna was struck by the lack of progress on climate over the eleven years since Al Gore produced *An Inconvenient Truth*. He turned his volunteer efforts to the climate crisis. Around the same time, he retired from his work as an engineer in the food industry and launched a Finger Lakes chapter of Gore's Climate Reality Project. Hirasuna's chapter has grown to eighty members over the past two years. Just last night I attended a webinar on converting methane from dairy farms to energy, one of many Climate Reality activities that Hirasuna organizes each month. He also volunteers with a group that monitors Ithaca's progress on our Green New Deal scorecard. Now that Ithaca has launched a public-private partnership to weatherize all city buildings,[54] Hirasuna is hopeful that our local innovations will spread throughout New York State and beyond.

Hirasuna joined the MIT alumni group for climate action, where he helps people use En-ROADS, an online tool that demonstrates the effect of policy changes on global warming. He mobilizes fellow alumni to take action and to advocate for climate policy with New York State legislators. Hirasuna also works on legislation with the Sierra Club's Atlantic Chapter and with the New York State Coalition of Climate Reality Chapters. And he has returned to Tompkins County Cooperative Extension, where he is on the Environmental Program Advisory Committee and again serves on the board, and is trained as an energy navigator who helps people save energy in their homes. Apart from his climate work, Hirasuna is in his third term as president of a local Toastmasters' club where he helps people improve their presentation skills. He credits his Toastmasters experience as being an essential part of his leadership of the local Climate Reality chapter. And he continues to play trumpet with our community orchestra and band.

Hirasuna is soft-spoken. But when I talked with him about his volunteerism, his enthusiasm for his climate work, and for all that he is learning, was evident from his constant smile and the fast pace at which he was speaking. He didn't seem at all depressed by his climate work. When I asked why he persists as a climate volunteer, Hirasuna responded, "Because things are not getting better. I would be happy if I didn't have to be an advocate." In the meantime, he achieves small successes, such as scoring meetings with congressional staffers, sending letters to representatives, and seeing how many people are hopping on the weekly actions he sends out to members of his Climate Reality chapter. Hirasuna recognizes that real success comes with legislation and is especially encouraged by recent progress in New York State as it works to meet the provisions of the Climate Leadership and Community Protection Act enacted in 2019.

Mobilization

> **Does the 3.5% rule apply to campaigns that aren't aimed at major results like removing a national leader or achieving independence—say, campaigns for climate action or against local governments, corporations, or schools? No one knows.**
>
> —Erica Chenoweth, political scientist, *Civil Resistance*

How many people does it take like Daneri or Hirasuna—or like you and me—to bring down a government, or at the very least an economic system, based on fossil fuels? After studying nearly four hundred resistance campaigns from 1945 to 2014, the political scientist Erica Chenoweth (who goes by they/them pronouns) concluded that no revolution has failed to overthrow a government once 3.5 percent of the population actively participated in a mass demonstration or other observable form of mass noncooperation. Although Extinction Rebellion and other climate organizations have cited Chenoweth's "3.5 percent rule,"

Chenoweth is unwilling to go out on a limb and predict what will happen in the future based on what they observed about the past. Chenowith also points out that the revolutions they studied were likely preceded by a long period of building public sympathy and support. Thus, the 3.5 percent of the population who were mobilized activists may have represented the views of 80 percent of the population.[55]

Regardless of whether the number needed to push radical change is exactly 3.5 percent, the work of Chenoweth and others demonstrates the importance of grassroots mobilization across generations through groups such as the youth climate organization the Sunrise Movement and Elders Climate Action as well as 350.org and its recent offshoot for older volunteers, Third Act. These organizations seek to change public opinion and build support for the climate justice cause—that is, to alter what is politically possible. This in turn sets the stage for their more active members to work with congressional or parliamentary representatives to help draft and implement new policies.[56]

Consumer and Protest Movements

Protests, graffiti-strewn monuments, and even taking over public parks come to mind when we envision movements such as Occupy Wall Street or Black Lives Matters. We can think of these political acts as being part of one type of social movement—that is, protest movements. Certainly the climate movement is a protest movement—witness the school strikes started by Greta Thunberg, which led to a series of global demonstrations involving millions of student protesters, the Friday afternoon vigils held by Jane Fonda and fellow elders in front of the US Capitol, and the disruptive Extinction Rebellion die-ins in London and other cities around the globe.

Yet the climate movement also has elements of a second type of social movement that focuses on lifestyles and uses consensus strategies.[57] Ethical consumption and political consumerism (see chapter 2) are lifestyle or consensus movements, where consumer practices present an alternative to unchecked capitalism and industrialization of food and other production systems.[58] Participation in lifestyle practices with friends and neighbors can create the connections that are essential to mobilizing people for further actions (including protest actions) and to sustaining a movement during periods less favorable to protest.[59] By growing their own food in community gardens, buying from local farmers, composting organic wastes, and patronizing cafés that serve sustainably grown coffee, lifestyle activists show us what our world can be through example, rather than protest. Perhaps most important, consumer movements can offer alternative "frames" or rationales for climate action, including for politicians who may want desperately to address the climate crisis but don't perceive climate as a top priority for their voting constituents. Consumer movements that frame plant-rich diet as a pathway to good health or not wasting food as a means to save money help "green-shy" politicians promote a climate agenda without focusing on climate per se.[60]

In contrast to consumer movements, political movements often directly challenge the political and economic system. "Defund the police!" or "Tax the rich!" they shout rather than quietly enlisting friends in composting or volunteering at a local food-donation center. Many social movements, as well as the organizations that support them, lie somewhere along the continuum of celebrating community and challenging governments. They combine events such as local harvest festivals with protests and boycotts. Their calls for social change reflect hybrid or "blended social action."[61]

Climate justice is one such movement that spans protest and consumer practices. The performative dumpster divers trespass on grocery store properties to challenge more powerful businesses and to protest the food system's wastefulness (see chapter 2). They also demonstrate an alternative, more sustainable, and equitable lifestyle by sharing the recovered food with friends and the unhoused. Capturing this "between space" in social movements, the political scientists Donatella della Porta and Mario Diani write, "In many cases, it is simply an issue of individual consumer behavior, no different from other fashion phenomena. In other cases, however, lifestyle becomes the stake in conflicts. . . . Youth movements and other oppositional countercultures provide examples of how individual lifestyle may take up an antagonistic character."[62]

Successful social movements promulgate a vision of an alternative to the existing system that appeals to a broad coalition.[63] Hence, we see the Green New Deal suggesting not just that we "save the planet" but also that saving the planet is really about saving people. This vision, promulgated by the Sunrise Movement and others, evokes clean energy replacing sooty coal and polluting oil, new jobs in the green energy sector, walkable cities, and a more even distribution of wealth. To achieve this vision, we will need to dismantle existing ways of doing business such as government subsidies to fossil fuel production; policies that sustain energy-intensive transportation, food, and energy; property and water rights that are incompatible with the new climate reality; and the influence granted to corporate lobbyists.[64] To achieve this vision we will also need grassroots organizations and their volunteers pressuring government to act.

Which Tipping Point Will We Reach?

> **Here's to the brilliant work of climate activists everywhere who are, together, creating a social tipping point for a fossil-fuel-free future.**
>
> —Kate Raworth, author of *Donut Economics*, Twitter

In spring 2021 *New York Times* climate correspondent Somini Sengupta wrote an article with the optimistic headline "Big Setbacks Propel Oil Giants toward a 'Tipping Point.'" She extolled the impressive victories against big oil of a nun, an environmental lawyer, pension fund executives, and the world's largest asset manager. But, she claimed, "the most dramatic turning point came in the Netherlands, where a court instructed Royal Dutch Shell, the largest private oil trader in the world and by far the largest company in the Netherlands itself, that it must sharply cut greenhouse gas emissions from all its global operations this decade. It was the first time a court ordered a private company to, in effect, change its business practice on climate grounds." This and other victories were opening new "battle fronts in the climate front," Sengupta enthused.[65]

Five months later, the American Petroleum Institute was buying Facebook ads targeting congressmembers backing climate legislation,[66] and Apple, Microsoft, Amazon, Disney, and the major pharmaceuticals, despite their promises to combat climate change, were backing groups actively undermining the first major US climate bill in over a decade.[67] Adding to the corporate anticlimate campaign, the fossil fuel industry sent more representatives to the Glasgow COP26 climate meetings than any single country delegation.[68] And Royal Dutch Shell announced its plan to move its headquarters from the Netherlands to the United

Kingdom, purportedly to take advantage of a lower tax rate and to better position its transition to a cleaner energy business.[69]

Repeatedly we hear scientists question whether recent unprecedented fires, droughts, hurricanes, floods, ice melting, heat waves, and now "bomb cyclones," alongside massive plant and animal extinctions, demonstrate that Earth has already reached an environmental tipping point. They suggest that we may have already lost the battle in our fight against climate change. Have we truly crossed the boundaries of life as we know it, with no going back?[70]

Although we can't know the answer, some social scientists, more optimistically, have proposed social tipping points that, if reached soon, could lead us to a more favorable future. They point to the relatively small interventions that spur "contagion," or rapid spread of new technologies, norms, behaviors, and structural change. A government removing fossil fuel subsidies or building a carbon-neutral city or mass protests that change norms and behaviors can become such "tipping point interventions."[71]

A long period of slower change generally precedes social tipping points. As I drafted this chapter, Donald Trump was president of the United States, and it seemed as if any efforts to promote climate legislation were futile. Who is in office and what climate legislation they support is even more important than lobbying

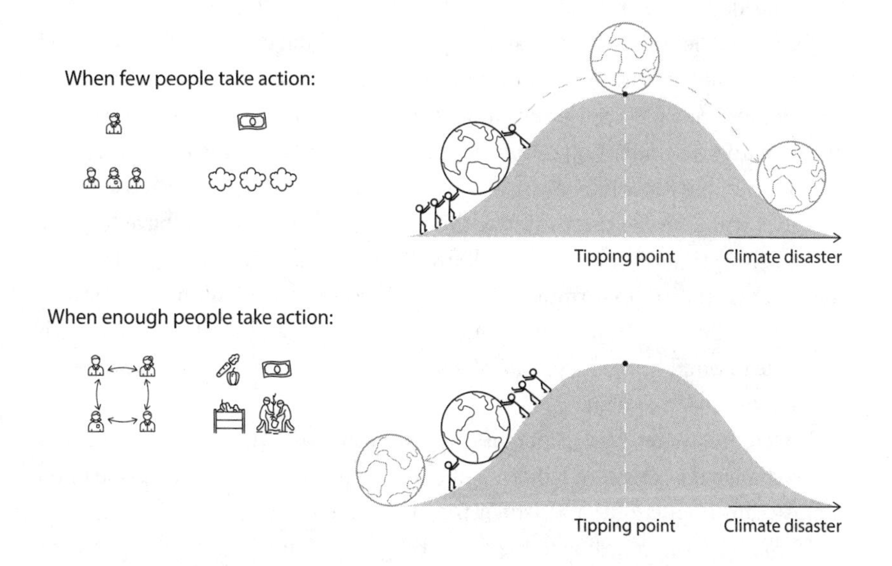

We may avoid tipping over into a world with more and more disasters if people, organizations, and governments adopt sustainable practices. Diagrams by Xiaoyi Zhu.

and special-interest money in determining the likelihood that climate advocacy will be successful.[72] Given that Trump claimed climate change was a hoax, you might think climate organizations such as Elders Climate Action and the Climate Reality Project would have been in a period of inactivity during his presidency.[73]

Yet climate activism, particularly at the local and state level, was gaining momentum. The city of Ithaca raised $100 million from private investors over just three months to fund loans for upgrading buildings to reduce emissions.[74] States such as California, whose governor declared that no new fossil-fuel vehicles would be sold after 2035, have also been leading the charge. And new types of businesses, such as benefit corporations (b-corps), are emerging with sustainability rather than profitability as their primary goal.[75]

Inspired by the charismatic Alexandria Ocasio-Cortez and young activists from the Sunrise Movement, and by their vision of a Green New Deal, climate and justice organizations in the United States have also been busy formulating plans and policies. Policy documents such as the Vision for Equitable Climate Change,[76] THRIVE (Transform, Heal, and Renew by Investing in a Vibrant Economy),[77] and Action for Climate Empowerment,[78] as well as President Joe Biden's Plan for a Clean Energy Revolution and Environmental Justice,[79] are being fleshed out and readied to be considered by Congress and government agencies. Yet, given the climate obfuscators in powerful positions and fossil fuel industry opposition in the United States and globally, whether we pass the legislation needed to reach an emissions reduction tipping point remains to be seen.

Still, hope lies in action—in avoiding environmental despair.[80] If we do reach a tipping point into rapidly drawing down greenhouse emissions, it will be in part because our lifestyle actions, donations, and volunteer activism have been building up over time. With more and more people eating plant-rich meals, support for policies to switch government subsidies from meat to plant-based commodities will no longer seem out of the question. Volunteers will have built the social connections and trust to work together. They also will have worked with their representatives to formulate policies such as those laid out in the Vision for Equitable Climate Action. And climate organizations and their volunteers will continue to build bridges with social justice groups, accelerating the work of the climate justice movement.

Yet let's not forget the role of mass protests, which are desperately needed if we are to change the "Overton window" of climate policies that are acceptable to the mainstream population, and which thereby influence government response to the climate crisis. According to the activist organizer Paul Engler, "major changes often come in punctuated bursts, and these bursts are most likely to occur after outbreaks of highly visible protest activity."[81] Volunteers who previously may have been satisfied with lowering meat consumption and food waste increasingly will be drawn on to participate in protests, including in civil disobedience. Upon turning sixty, the climate writer and activist Bill McKibben launched Third Act, a climate advocacy organization for people in their "third act" of life. During a Zoom call announcing the new organization, participants were asked to respond to a poll asking about their interest in civil disobedience training. I responded that I was "interested in learning more." I hope that Third Act along with the Sunrise Movement and others will help mobilize volunteers like me—that they will engage 3.5 percent of the US population (eleven and a half million) or whatever it takes—to participate in those acts of mass mobilization that historically have toppled governments and that may transform our energy, economic, and food systems to end the current climate devastation.[82]

• • •

VOLUNTEER: Don'ts and Dos

1. Do find a volunteer role that suits you with an organization that is influencing climate policy. The Climate Reality Project may be a good place to start because it has a lot of local chapters where you can find people nearby to work with. Younger people might want to consider Sunrise Movement, whereas older folks can look into Elders Climate Action or Third Act.

2. Don't wait as long as I did to find your volunteer role. The climate can no longer wait for us to act.

3. Do take the time to read the emails and take the actions that the organizations you belong to recommend.

4. Do download the Climate Action Now app. It has a large repertoire of quick actions you can take to influence policy. Your organization can also post its own actions on the app for others to engage in.

5. Do connect with other climate-concerned citizens and consider joining protests and even civil disobedience.

5

RESPONSIBILITY

The gospel of individual responsibility always plays well with American audiences.

—Amy Westervelt, journalist, *Drilled* podcast

By now it's common knowledge that for over fifty years the fossil fuel industry worked frantically to obfuscate the truth about climate science, even as its own scientists reported that greenhouse gas emissions were rising to dangerous levels.[1] What is less well known is that the oil companies launched a parallel disinformation campaign to escalate the war against climate policy. Starting in the early 1990s, the American Petroleum Institute commissioned economists from Charles River Associates to develop predictions about the economic impacts of regulating emissions. Not surprisingly, given where their money was coming from, the economists predicted dire economic consequences should the government regulate emissions. At the same time, they failed to consider the financial fallout of runaway climate change. News outlets such as the *New York Times*, *USA Today*, and CNN promulgated the faulty economic predictions without acknowledging that the fossil fuel industry had paid for the misleading numbers.[2]

The economists' predictions—tainted by industry dollars—have repeatedly influenced climate legislation. President George W. Bush cited the Charles River Associates findings in withdrawing from the Kyoto Protocol, as did President Trump when he exited the Paris Agreement.[3] And when Senators John McCain and Joe Liebermann introduced a climate emissions cap-and-trade bill back in 2005, none other than Senator Jim Inhofe, infamous for throwing a snowball inside Congress to prove that the climate was not heating up, had this to say:

> Enacting the McCain-Lieberman bill would cost, according to Charles River Associates, the U.S. economy $507 billion in 2020, $545 billion in 2025. . . . Under Kyoto, for the average family of four in America, it would cost them $2,700 a year. This bill will only cost them $2,000 a year. So maybe that isn't quite as bad as it would have been otherwise.
>
> The bottom line: It is very expensive. And that is not just Senator INHOFE talking. We are quoting CRA [Charles River Associates], which is the recognized authority. [4]

Inhofe failed to calculate the price—in dollars and human suffering—of not acting on climate change. Although the annual costs of climate disasters have not yet reached the biased economic projections Inhofe attached to the McCain-Lieberman Bill, they are creeping upward each year. In late December 2021, Christian Aid calculated the estimated costs of major climate disasters in the United States at $23 billion for the Texas freeze and $65 billion for Hurricane Ida but did not include figures for the California fires or the Pacific Northwest heat wave.[5] Just days after the report was released, nearly one thousand homes in suburban Denver were burned to the ground, adding immense costs and suffering.

Over time, the Charles River Associates economist Paul Bernstein became increasingly concerned about climate change and later wrote, "I think the API [American Petroleum Institute] knew that if they had Charles River Associates run these models, it would produce the results the API wanted, namely that it would show a cost to [climate mitigation] policy."[6]

As it was funding scientists to obfuscate the science and economists to obfuscate the financial outfall of climate change, the fossil fuel industry continued its efforts to convince the public that we needed its polluting products, that only oil and gas could bring us economic prosperity and happiness. Consequently, it's not surprising that climate activists blame the fossil fuel companies and become incensed when individual citizens are called on to solve the climate crisis. It's also understandable that *The Guardian* journalist and *Drilled* podcast producer Amy Westervelt exhorts us to ignore the oil companies' insidious "gospel of individual responsibility."[7]

At the core of Westervelt's argument against this individual responsibility gospel is that big oil is to blame and therefore must accept responsibility. I, an individual citizen, am simply an unwilling consumer of what is available in terms of

food, transportation, energy, and other commodities. If the government worked with Elon Musk and the CEO of Ford to install a nationwide efficient system for recharging electric cars or if my power company made it easy to switch to renewable energy, I would happily move away from my gasoline-powered car and oil hot-water heater.

Yet, in 2020, the Count Us In campaign was launched to inspire one billion individuals to take a climate action. Those who join the campaign can choose consumer actions, such as eating more plants and dialing down their thermostat. They can also choose collective actions aimed at systemic change, such as talking to politicians and investing their money responsibly. Count Us In claims that the billion individual actions of people around the world will have a significant impact on carbon emissions. They also state that the actions will send a message to companies and governments to take action themselves on climate change.[8] Not everyone agrees. The youth climate activist Joel Lev-Tov called the campaign "absolutely disgusting. It's blaming the consumers for their choices instead of the fossil fuel companies and big business who are emitting more CO_2 than I could ever emit in my lifetime in a few hours. While I appreciate the intention behind their campaign . . . individual change won't help us solve the climate crisis."[9]

The more I read about the mendacity and nefariousness of the fossil fuel industry, the more appalled I became. If it had only acted differently, people in New York City would not be breathing smoke from the California fires, and our climate fellow from Fiji would not be watching much of his country drown. But efforts to hold the fossil fuel companies accountable take years to wend their way through Congress and the courts.[10] We can't afford to wait. So, I fall back on eating less meat and cutting down on the food I waste, donating to Femme International, and volunteering for Elders Climate Action.

Yet I am troubled by an awareness of how slight my contribution is. Those who call for structural changes to our fossil-fuel-dependent economy are right. Furthermore, I understand that fossil fuel executives and other "polluter elites" have more power than I do. They not only have access to financial decisions and policymakers but also are members of powerful elite networks, which they mobilize for their pro-fossil-fuel cause.[11] And then there's their luxury consumption, which emits vast amounts of carbon and spurs others to copy their lavish behavior.[12]

Arguments about who is to blame and who is responsible for the climate crisis, and what that means for responsible actors going forward, can occupy a lot of mental energy. Disagreements can derail climate action. Fortunately, powerful thinkers have pondered and debated questions of blame and personal responsibility for years, and their writings can help us sort out ways to move forward in the climate crisis. Two such philosophers are Iris Marion Young, who is known for her social connection model of responsibility, and Robin Zheng, who looked at responsibility through the notion of "role-ideals."

Regardless of Young and Zheng's mix of terms and approaches, they both would agree on one thing: *forward-looking responsibility*. This means we should place an emphasis on identifying who is responsible to take action moving forward rather than looking backward at who caused the crisis.[13] They might ask, Regardless of who is to blame, what good does blaming do to solve the climate crisis?

What Does Blame Achieve?

Finding someone to blame for perceived ills has been at the forefront of US policies and politics for years. This is particularly the case in the struggle to advance economic and racial equity. Take, for example, the reforms of the New Deal, including social security and welfare, which caused a backlash where officials blamed the poor for their problems. Later, President Ronald Reagan's derision of "welfare queens," who drove Cadillacs and donned fancy furs while living on the dole, permeated the airwaves. Such stereotyping and blaming led Reagan to victory and President Bill Clinton to subsequently reform the welfare system.[14] The reforms blamed the poor for being poor, penalized them by imposing unattainable work and marriage requirements for receiving public assistance, and ignored systems of racial and economic injustice.[15] Following the 2020 Black Lives Matter protests, we saw a continuation of these arguments on the political right. Blacks were to blame for crime in their neighborhoods, for riots that burned down businesses, and for the way their behaviors incited police to kill them. Those on the left also sought someone to blame—the police, President Trump, and systems of racial and economic injustice.

For a long time, I thought I shared no blame for racial injustice, given that I support Democratic candidates and my ancestors were not slaveholders. My mother's family came over on the Mayflower to Massachusetts, and my father was an Austrian-born ethnic Jew who escaped the Nazi massacre by immigrating to the United States. But then I discovered that when my parents bought their first house in Maryland in the 1950s, they signed a racial covenant, promising never to resell their house to "Negroes." Such covenants were pervasive at the time.[16] My parents accumulated wealth through owning the house, and part of that wealth was passed on to me. African Americans were routinely denied the opportunity to buy homes during most of the twentieth century, which led to generational poverty and massive inequality.

Aren't I, then, a part of the system of racial injustice in the United States? Does it matter that my parents would have preferred not to sign that covenant? Although I benefited financially from racial covenants, philosophers might question whether it is appropriate to blame me, as so many people like my parents were forced into signing racial covenants. Blame is more justified in situations where people exercise their free will to enter into agreements. Similarly, am I to blame for climate injustice if I still drive a gasoline-powered car and fly in airplanes, albeit reluctantly or even involuntarily?[17]

Seeing how ubiquitous housing covenants were in the 1950s, some might claim that the real estate industry and government that supported housing discrimination are to blame. Philosopher Iris Marion Young is more nuanced. She would argue that even though I alongside the fossil fuel companies, real estate industry, and government bear *responsibility* as participants in systems of injustice, *blaming* them (or blaming me for that matter) will be unproductive.

Young divorces responsibility from blame. One problem with blaming specific actors is that it lets everyone else off the hook. If the poor are to blame for being poor, then I, a middle-class professor, don't need to engage in addressing poverty. If only criminals are to blame for crime and only violent police to blame for police violence, then I, as a citizen who is neither a criminal nor a cop, am absolved of responsibility. But, Young urges, examine the bigger system of injustice. Because we are all part of economic and social systems, we all play a role in whatever injustice is perpetrated through these systems.

Blaming also deflects attention from the background conditions in which individuals and industry operate. Companies may feel pressure to maximize their profits. Top executives may feel compelled to conform with the norms of their elite peers to earn excessive salaries and demonstrate their wealth through elaborate mansions and expensive yachts. Importantly, blaming the companies puts the responsibility, and perhaps the power, for change only on the elites. This can lead to narrow solutions that reinforce the elites' powerful position and prevent equitable systemic change. In other words, just because the companies might deserve the blame, that doesn't mean blaming them will be the best response if we want to equitably solve the problems they created.[18]

Another way that blaming is unproductive is that it induces defensiveness. Look at how hard the oil companies are fighting back against being blamed for the climate crisis. Similarly, all kinds of strong arguments spring to mind for why the covenant my parents signed when they bought their house in the 1950s is not my fault.

Feeling defensive is just one way blame can hinder a person's ability to help solve climate and other problems. I was recently interviewed by a graduate student researching elders' views on climate change. One of his questions was whether I felt like I was to blame, as a baby boomer whose generation had seen greenhouse gas emissions skyrocket during my lifetime and whose behaviors, including not advocating to hold government and industry accountable, contributed to those harmful emissions. I responded, no, that I had not really considered blaming myself. For one, I felt I was doing more to address the climate crisis than most people around me. I, like so many others, also lacked good alternatives to gasoline-powered cars. But, most important, blaming myself, as I do every day for other failures in my life, is harmful for my mental health and not particularly productive. This sort of backward thinking says little about what I can do to rectify the harm.[19]

Closely connected to notions of blame is liability, which identifies legally or morally responsible actors for the purpose of sanctioning, punishing, or extracting compensation. Whereas liability is necessary in a legal system, Young claims that it, like blame, fails to address structural injustice. BP paid billions of dollars after being held legally liable for the Deepwater Horizon oil spill. But how much did that payment change the systemic climate injustices perpetuated by oil companies and lax government regulation? Following the massive oil spill, BP stonewalled Congress by withholding information,[20] and it poured $200 million into its Beyond Petroleum rebranding campaign,[21] while actively seeking to exploit new sources of oil. It eventually recovered from the billions in fines to once again become a major global player in the fossil fuel industry.[22] As part of its Beyond Petroleum campaign, BP began to tout its commitment to sustainability, which was more greenwashing than reflective of any systemic change within the company. Ten years after the largest oil spill in US history, we still depended heavily on fossil fuels for energy and still lacked regulations to prevent future ocean oil spills.[23] When the next major storm, Hurricane Ida, hit New Orleans in 2021, oil once again gushed out into the Gulf of Mexico, while in California an oil pipeline broke and a massive spill bespoiled birds and beaches.

What Can Shaming Achieve?

On October 28, 2021, Congress held a heated virtual hearing with the CEOs of Exxon Mobil, BP, Chevron, and Shell, alongside their lobbyist friends from the American Petroleum Institute and the US Chamber of Commerce. According to Congresswoman Carolyn Maloney, who chaired the hearing, the goal of the hearings and subsequent subpoenas was to "get to the bottom of the oil industry's disinformation campaign."[24] Maloney continued: "For the first time, top fossil fuel executives are testifying together before Congress, under oath, about the industry's role in causing climate change—and their efforts to cover it up. For far too long, Big Oil has escaped accountability for its central role in bringing our planet to the brink of a climate catastrophe."[25]

Throughout the hearings, members of Congress heaped on the blame in an effort to get the companies to "fess up" or "pay up" for their sins.[26] Perhaps hoping to recreate the high-profile tobacco industry hearings of the 1990s, where cigarette company executives shamed themselves by falsely testifying that smoking was not addictive, Democrats tried to shame the oil industry execs for misleading the public while lining their pockets.[27]

Pouring different quantities of M&Ms into jars as a prop, Representative Katie Porter compared the $20 billion Shell plans to spend on oil, gas, chemicals, and marketing to the $2–$3 billion allocated to renewable energy. Shell CEO Gretchen Watkins responded with a thinly veiled attempt to shift the blame to consumers: "There needs to be both a demand and a supply of clean energy, which is why we're working very closely with our customers so that that demand increases over time."

Porter shot back: "To me, this does not look like an adequate response to one of the defining challenges of our time. This is greenwashing! Shell is trying to fool people into thinking it's addressing the climate crisis when what it's actually doing is to continue to put money into fossil fuels."[28]

When Representative Jamie Raskin asked the executives whether they "accept that the First Amendment does not protect fraudulent commercial speech," all four disclaimed any expertise on matters of the Constitution. And when asked by Representative Maloney to respond "yes or no" to a question about whether climate change presented an existential crisis to humanity, all six industry leaders remained silent.[29]

Silence and obfuscation are perhaps expected when CEOs are blamed or shamed publicly—especially when facts and history are not on their side. There was also ample evidence of greenwashing from the CEOs and their lobbying partners as they tried to shift the focus away from what caused the climate crisis to the oil industry's efforts to find solutions. Philosophers such as Iris Marion Young would point to tactics that evoke blame and backward-looking responsibility as being ineffective. Certainly Congress's tactics did not immediately elicit a helpful response.

The goal of the hearings seemingly was to put the companies on trial rather than work toward solutions. The purpose was to look backward at who caused the crisis rather than forward to who would solve it. Perhaps away from the spotlight of public hearings, Congress will work to find some exemplary executives to start working toward solutions and follow the advice of Robin Zheng: "Rather than accusing a person of complicity in structural injustice, which is likely to provoke defensiveness, one can instead cite exemplary role models, organizational mission statements, or other discourses that serve to prime a person's role-ideals."[30]

Yet shaming has its purpose. For shaming campaigns to be successful, consumers need reliable information about the corporations' behaviors, which can be provided by civil society organizations that review company documents.[31] In the case of the oil companies' misdeeds, we know about their nefarious actions from academic and journalistic research into their documents and public communications.[32]

Jennifer Jacquet has written an entire book on the purposes of shaming, which she feels is a crucial tool for holding bad actors responsible. She has even outlined "the seven habits of highly effective shaming," which I illustrate using the oil executives congressional hearing:

1. The bad actors must be concerned about their bad behavior. (The oil giants should be concerned about their contributions to greenhouse gas emissions.)
2. The gap between desired behavior and what the bad actors did should be large. (The oil giants should be transparent about their products' impact on the planet.)
3. Formal punishment is lacking. (Currently, the government does not have a means to sanction the oil industry's behavior, although that may be changing as lawsuits against oil companies wend their way through the courts.)
4. The bad actor should be sensitive to the group that is shaming. (Oil companies care what Congress thinks.)
5. The group that is shaming is trusted by the audience. (At least a portion of the American public trusts Congress, which is, after all, democratically elected.)
6. Shaming should be directed where potential benefits are large. (Oil giants have the ability to pay vast sums to compensate for their behavior)
7. Shaming should be carefully implemented. (Congress has considered whether their public shaming of the oil industry via an open hearing will lead to any changes.)[33]

It remains to be seen how effective the congressional shaming event will be, although I suspect it is just one phase of a much larger effort to hold the oil companies responsible.

Interestingly, the most effective industry-shaming campaigns may not be those orchestrated by elected representatives but by consumers. In at least three cases, shaming has led large multinational companies to accept the criticism and redress their past wrongs. In the 1990s massive public disapproval of Nike's exploitative labor practices led to loss of profits and reputation and to Nike reforming its supply chain. When Shell proposed sinking a defunct oil platform in the North Sea, Greenpeace launched a powerful shaming campaign resulting in a widespread boycott of Shell gas stations in Northern Europe. Shell abandoned its plan. And in 2010 Greenpeace posted on YouTube a sixty-second video of a bored office worker opening up a KitKat wrapper and finding, instead of a chocolate bar, an orangutan finger—fur, bones, and all—which he nonchalantly eats while blood drips down his chin and colleagues look on in horror. The video was meant to shame Nestlé into stopping its clearing of orangutan forest habitats for palm oil plantations and forced the company to develop a timetable for cleaning up its palm oil supply chain.[34]

Forward-Looking Responsibility

> **Determined and enlightened government can make a difference. . . . As we know this will only happen if there is public pressure.**
>
> —Dario Kenner, author, *The Polluter Elite and the Challenge of Rapid Transition*

Those blaming the oil companies for fossil fuel pollution, or my parents for signing racial covenants, are engaging in what is called backward-looking collective responsibility. Backward-looking responsibility focuses on who *caused* an existing unfortunate state of affairs. In contrast, forward-looking collective responsibility asks the questions, Who can bring about a better state of affairs, or who can help to solve the problem?[35]

How do we decide who has responsibility for rectifying a situation? First, we can consider practicality: the agent that created the harm may or may not be best situated to improve things. The climate philosopher Elizabeth Cripps posits that two groups of people have a particular duty to mitigate climate change: the able (those who because of their wealth can contribute to climate action without great cost to themselves) and the polluters (those who contribute more than their fair share of emissions).[36] The oil industry is both able and a polluter, but its continued insistence on a business model based on burning fossil fuels argues against it being the best actor out front addressing the climate crisis. Government as well as many nonprofit and grassroots groups and their volunteers are also both able and have contributed their fair share to pollution and are well positioned to be out front finding climate solutions. Although fairness is important and eventually the fossil fuel companies will hopefully be forced to pay damages, fairly distributing responsibility is limited by the need to find immediate practical solutions to the climate crisis.[37] To put it bluntly, the best solutions might be those in which some bear responsibilities that should, ideally and ethically, be borne by others.[38]

To go beyond responsible actors and guide actual actions, Cripps examines several types of moral duties to address climate change and lands on "cooperative promotional duties" as primary.[39] "Cooperative" refers to the importance of taking actions with others, whereas "promotional" indicates that such actions should aim for effective global-level progress on addressing climate change. Cripps eschews the idea that our primary moral duty is to cut our own emissions and instead feels that progress will come in the form of collective action.[40]

Social Connection

In her posthumously published book *Responsibility for Justice*, Iris Marion Young explores notions of blame, liability, and responsibility in relation to global systems of injustice.[41] She proposes the *social connection* model of responsibility and uses the textile industry, whose sweatshop workers in poor countries feed richer countries' hunger for cheap clothes or even fast fashion, to illustrate her points.

According to Young, everyone who plays a role in the global apparel system—whether they be a billionaire business magnate, small factory owner, consumer, or sweatshop laborer—is connected to that system. And by dint of their connection to the system, each actor bears responsibility for the system's injustice. Each of us, however, bears a different level of responsibility, depending in large part on our ability to address a system's injustices moving forward.

Young further claims that injustice in a production system comes about not because each actor wants to inflict harm but rather because the system constrains or even dictates how we act. No actor is necessarily evil within their particular context or network. The elite business CEO is acting according to

Production system

Iris Marion Young's social connectedness model shows how all actors are connected and bear some responsibility for injustices in a food, textile, or other production system. Diagram by Xiaoyi Zhu.

the norms of elite business executives. They are also constrained by their company's by-laws, which likely prioritize earning profits for shareholders over corporate responsibility and good citizenship. The factory owner employs textile workers as part of normal everyday business, and were they to increase wages, their factory might be unable to compete with other factories. The consumer is buying clothes that reflect the choices of their friends or social class, and if they were all to boycott a clothing brand, textile workers in Bangladesh could lose their jobs.

Young claims that by focusing only on the wrongdoing of particular actors such as the oil executives or by blaming them, we sidestep the structural injustice that occurs when all of us simply do what we normally do, guided by accepted rules and norms. Of course, as we follow norms, there is undoubtedly something, however small, we could do to challenge a system that exploits workers, pollutes streams, and emits damaging greenhouse gases. All of us play a part in this injustice, despite the fact that we live in different places and have different roles as producers, consumers, and members of society. Furthermore, an individual is not only responsible for the climate emissions they cause directly, as measured by the latest carbon footprint calculator. They are also responsible for being an active participant in the apparel, food, and other systems that make our lives dependent on fossil fuels.[42] This kind of responsibility comprises more than what we consume in our everyday lives. It also includes the political realm. Young asks us to "collectively transform the background structural injustice within which we live our lives."[43]

Social injustice is thus created by the everyday acts of millions of actors who are connected by their participation in food, apparel, or other systems. Most of us are not intentionally causing harm. And we may not be aware of the harm we are causing by our normal routines or practices. We may even want to try to make things better. But by the nature of our everyday acts, we still contribute to the unjust system. Thus, the social connection model reveals our connection not only to other actors but also to structural injustice—that is, "*the reproduction of unjust structures through individuals' actions*."[44]

Young's work offers an alternative to the debate we began with, which pitches individual against corporate responsibility. No party is absolved of responsibility, but no single actor is solely responsible. Instead, we shift our focus from individual actors to the systems and processes that are entailed in our unjust production systems, and we strive to understand how our participation in these systems enables them to function and persist and what actions we can take to change them.

Levels of Responsibility Matter

You might object to the idea that you, a consumer, should be put into the same basket of responsibility as the multinational corporate CEO. Young has a response: we don't all share the same *level* of responsibility. And different levels of responsibility call for different levels of response.

According to Young, we should weigh each actor's level of responsibility using four criteria: power, privilege, interest, and collective ability.[45] The farmer, corporate executive, grocery store manager, and consumer should ask themselves such questions as:

Power: Which aspects of climate injustice do I have the most power to change? How might I use my own power to exert pressure on the more powerful agents, such as food companies and government?

Privilege: How am I benefiting from the current system? For example, middle-class consumers living in a democracy benefit from a vast array of affordable food choices and have opportunities to protest. Compared to poorer citizens or those living in an autocracy or dictatorship, wealthier consumers living in a democracy may have better access to alternative, climate-friendly options such as electric vehicles and heat pumps, as well as more opportunities to influence government. How might I use my privilege to adjust my climate consumer and advocacy practices?

Interest: How might I address those issues in which I have a particular stake or interest?

Collective ability: Given my role in society (e.g., consumer, banker, farmer) and my membership in organizations and groups (e.g., church, sorority, CSA, chamber of commerce, climate organization), on which issues might I most effectively work collectively with others to alleviate injustice?

After reading Young's work, I weighed her four criteria while considering questions about how I might most effectively address climate injustice. The criteria, however, seemed too general to direct practical action. In fact, Young did not intend to guide specific actions. She wrote that countering injustice is less about what to do and more about an ongoing exploration of victims' needs and encouraging others to help address those needs in ways that reflect one's power, privilege, interest, and collective abilities.

Nevertheless, Young's ideas about social connection and responsibility are helpful. They empower people by suggesting alternatives to simply blaming ourselves *or* the big companies. Young would argue that as people who eat, we all bear partial responsibility for food system and related climate injustice. So do the other actors—grocers, restaurant chefs, food writers, multinational CEOs, laborers, and farmers—bear some responsibility, although our responsibility differs depending on our power, privilege, interest, and collective abilities.

The actors in Young's production systems are connected, but their connections are distant, unlike the close connections among family and friends needed to spread climate-friendly behaviors. Yet Young also recognizes the importance of being able to influence our close networks. She writes, "Political responsibility is not about doing something by myself, however, but about exhorting others to join me in collective action. When this occurs, and it occurs relatively infrequently, movement participants are often the most surprised at the transformative power they turn out to have."[46]

Role-Ideals

> **It is everyone's job to fight injustice because it is already their job to perform their roles well.**
>
> —Robin Zheng, philosopher, *What Is My Role in Changing the System?*

According to Young, I and others have responsibility, which is determined by our power, privilege, interest, and collective ability. But what can I, and my family and friends, do about this responsibility?

Philosopher Robin Zheng picks up where Young left off. Instead of focusing on the connectedness of a production system's actors, such as factory worker, banker, and consumer, she looks to our social roles, such as parent, teacher, student, club member, volunteer, and citizen.[47]

Zheng looks at questions of responsibility through "role-ideals." She claims that within any one role (e.g., teacher, father), individuals strive to do their best—to act according to the ideal they hold for that role.[48] Furthermore, our responsibility for structural injustice can be found in our social roles—that is, our roles as parents, colleagues, employers, and citizens—because these are the very roles that determine how we act within a particular social or institutional environment. Although we may not be able to attain our role-ideal—be the perfect mom or nurse or churchgoer—we still have those role-ideals in our minds and use them to guide our parenting, work, and faith-related decisions. We are motivated to act consistently with our role-ideals because our roles are part of our identity, foundational to who we are and how we act.[49] We also derive satisfaction from meeting our expectations for being a mom, a neighbor, a volunteer. No one wants to be labeled a bad mom, a bad teacher, a bad neighbor, or even a bad citizen or volunteer.

Harking back to Young, unjust systems are composed of connected social actors. But Zheng asks us to switch the focus from somewhat abstract actors—such as consumer—to roles that are truly part of our identity—such as

"In her role as a mother and investment banker, Megan was ideal. As a Rockette, less so."

churchgoer, volunteer, or sibling. By modifying how we act in each of our roles, we can change (however slightly) the overall system.[50]

It is through our roles that we have agency to change norms, rules, and institutions—what are collectively called structures. And it goes both ways. By changing rules and other structures, we enable people to make changes in how they perform their roles.[51] In his role as chair of Elders Climate Action's chapter in Ann Arbor, Michigan, Joe Ohren worked with his city government to implement a pilot curbside food-waste pickup system—the pilot effected small structural changes by providing bins and free pickup for households that separated out food waste. The structural changes made it easier for residents to recycle their food waste, changing how they were able to perform their roles as climate-concerned citizens. As a result of his efforts, Ohren's volunteer role subsequently shifted. He was invited onto the city's solid waste management committee to help advise on further changes in how Ann Arbor handles waste. Ohren's experience has parallels to my own when I started a travel climate offsets initiative in my department and was then invited onto the university's low-carbon travel working group. Trying to live up to one's role-ideal in one setting, such as an employee persuading colleagues to offset travel, may spill over to opportunities for new roles in other settings. In both Ohren's and my cases, the new opportunities put us closer to making systemic change in our city or workplace.[52]

Even if we care about a particular issue, such as climate change, we are unlikely to be able to perform perfectly in each of our roles. But, according to Zheng, "we're expected to do the best we can" as a teacher, a mom, a citizen, a volunteer.[53] We strive to meet our expectations about how we should act in any one role.

Importantly, doing our best entails *pushing the boundaries* of each of our roles.[54] For example, at work I decided that the climate crisis warranted pushing against the boundaries of the "professor-as-unbiased-purveyor-of-information" role. I developed new courses about network climate action and required students to conduct climate actions with their family and friend networks and with nonprofit or business partners. In doing so, I had to ignore the chair of our department curriculum committee, who told me doing activist projects with students was not consistent with our educator role. I also volunteered to lead the department seminar committee and asked each speaker to share not just their research but also their perspective on "crisis disciplines"—areas of research that address existential threats such as extinction and global warming and in which scientists should therefore go beyond business as usual to consider ethics and advocacy.[55] As a parent, I try to push boundaries by gently nudging my children toward lower-carbon lifestyles, while accepting their choices so they won't feel that I am "blaming" them for being bad. Perhaps where it's been hardest for me to push my boundaries is in my role as a volunteer advocate, where I can find a million excuses for avoiding actions that involve calling my representative, knocking on doors, or phone-banking potential environmental voters. But this shortcoming is being overcome by fun uses of technology, such as Zoom action parties, where instead of contacting senators and house members in the isolation of our homes, people socialize while sending advocacy letters to elected representatives from their phones.

Naturally, people will differ widely on what they consider to be the role-ideal for a teacher, parent, or volunteer. Here Zheng calls on us to reflect on our motives. Are we acting out of sincere belief in a particular strategy to promote justice? Or are we acting out of bad faith, ignorance motivated by an unjust worldview, ulterior motives such as greed and malice, or attitudes such as racism or classism?[56] Or are we simply doing things the way we have always done? To address injustice, we need to perform all our "roles with a *raised consciousness*."[57]

Might it be best to focus simply on the role in which we have the most power? Zheng responds that we should strive to do the best we can in all of our roles. Her argument: Striving in general to do better is something we do already as parents, teachers, or volunteers, and addressing injustice is part of striving to do better. We must simply add the criterion of promoting climate justice to criteria we are already considering.

As we strive to perform well and push the boundaries of our roles, Zheng offers several pointers to guide our actions. First, we should consider where we can act collectively to have the most impact. This may be through mobilizing our colleagues and friends to eat less meat or through volunteering with a climate or social justice organization and of course asking our friends to volunteer with us. If we are already a member of a climate nonprofit, expanding our collective action can mean being attuned to social justice organizations' interest in climate and then following up on that interest to help form climate justice coalitions.[58]

Second, we need to consider which actions have the greatest potential to change norms and policies—that is, the possibility for making structural change.

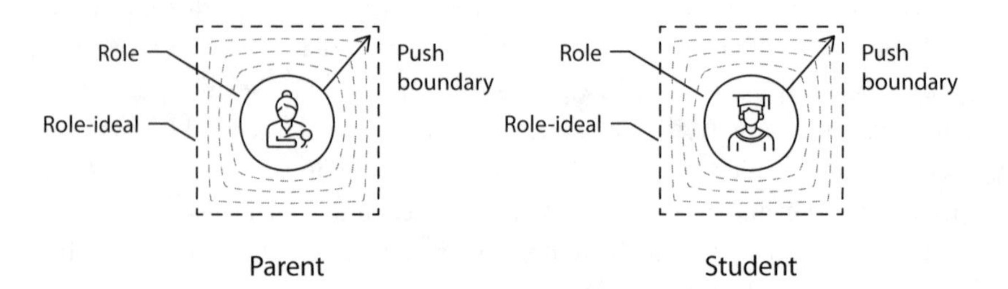

Robin Zheng asks us to push the boundaries in each of our roles as we strive to reach our role-ideals. Diagrams by Xiaoyi Zhu.

I can be a better shopper and reduce my family's food wastes, but I can also make suggestions to my grocer to put low-carbon foods in accessible locations or ask them to donate the store's food waste to a soup kitchen. Going to a protest may be one way to enact our responsibility, but unless the protest is preceded and followed by organizing activities, it may not "stick."[59] Volunteering with an organization that helps members engage in activities on an ongoing basis is more likely to affect norms and policies.

Additional criteria in choosing actions include whether the actions are visible to others, thus enabling the behavior and related norms to spread more readily. Harking back to my plalking habit (see introduction), I can pick up trash after dark while no one is looking or instead in plain sight of other walkers, thereby perhaps introducing them to plalking. Finally, one would want to consider ease and familiarity of a particular behavior and how efficient the behavior might be in achieving our climate justice goals, as well as personal preference or comfort with certain actions.

While not telling us exactly what we should do, Zheng's role-ideal model of responsibility provides a framework that is useful in everyday life. When we reflect on how to ideally perform our roles, we assess how we can promote justice as a mom, brother, carpenter, artist, banker, investor, volunteer, or any of our other roles. Because we are already performing these roles, the task is relatively small. We are already buying food for our family. What adjustments can we make to add more plant-rich meals and reduce the amount of food we throw out? We are already teaching classes. What small changes can we make to add climate justice to our curriculum? Or, as a carpenter or real estate agent, what can I do to make buildings more energy efficient or to steer homebuyers to greener homes?

Of course, when it comes to roles that are less intrinsic to our core sense of identity, not everyone is searching to do the best they can to fulfill a role-ideal. Case in point: the role of "consumer." Whereas we often describe ourselves as dads or New Yorkers or nurses, few of us would describe ourselves as consumers. In focus groups with consumers in the United Kingdom, researchers found that "sometimes when people talk about their roles as consumers they accept that they do have certain responsibilities; sometimes they make excuses for not doing more; but sometimes they make pertinent sounding justifications for not considering it their responsibility at all; and maybe, just maybe, if you listen hard enough, they might be asserting finite limits to how much they, as individuals, can be expected to be responsible for, and they might also be articulating justifiable skepticism towards the whole frame of 'responsibility' that is being addressed to them."[60] Perhaps had the participants been asked about roles more intrinsic to their identity than consumer, such as parent or neighbor, they would have been more motivated to take on responsibility.

What Can We Do Together?

> **I very much don't want someone who simply won't ever go march on the streets to be able to think that they are thereby off the hook. There are many, many other things they can be expected to do.**
>
> —Robin Zheng, philosopher, *Pea Soup*

If holding the fossil fuel industry and government accountable for their actions by joining a protest, signing on to a climate advocacy letter, supporting congressional representatives who hold hearings with oil company executives, or even joining a class action suit makes the most sense to you in your role as activist or someone with policy or legal expertise, then please go ahead and do so. But if trying out vegan or leftover meals with friends makes the most sense for you, given your power, privilege, interest, and collective abilities, then go for that. When it comes to addressing the climate crisis, none of us is off the hook, regardless of whether we are well suited for any one action. There's always something we can do.

We have limits on what we can handle at different stages and as circumstances change in life, but most of us can contribute something as parents, students, employees, churchgoers, club members, philanthropists, and volunteers. Although Zheng says that we should try to live up to all of our role-ideals, this may feel prohibitively overwhelming. It's fine to start with the role where we can most readily act according to our "highest standard,"[61] while keeping our eye on how we can influence those close to us to join in.

With Zheng's work in mind, I have started to define my own role-ideals. In each of my roles, I continuously remind myself to reflect on my ideals and how I can better live up to them. The first role that springs to mind is news consumer, given my compulsive daily habit of clicking on the latest headlines. Recently, while reading *The Guardian*, I was caught by the headline "The 20 Best Sandwich Recipes." Just three weeks earlier, *The Guardian* had announced that it was "doubling down on its climate commitments and launching further editorial and business initiatives one year after pledging to reach net zero greenhouse gas emissions by 2030."[62] So, I was appalled to see that thirteen of their twenty sandwiches featured meat or dairy and only one was totally vegan (a peanut butter and jelly sandwich, shown with the bread's crust removed and presumably discarded).[63] I made an online comment on the article suggesting they publish a new article with twenty sandwiches consistent with their paper's climate commitment. But after I clicked on "Submit," I couldn't find my entry among the over eight hundred comments extolling the tasty sandwiches. So, I emailed *The Guardian*, which responded that they would pass on my comment to their food writer.

Other small actions I've taken to move my news-consumer role toward something more ideal have include emailing podcasters, such as Preet Bharara and Charlie Sykes, asking them to improve their climate coverage. I also contacted two scientific journals to ask if they were reconsidering their policy of posting articles by Garrett Hardin without any comment on how his white supremacist views are now considered abhorrent. I even got a constructive response from the journal editors, who contacted the publisher. As in many attempts to change things, the response has been slow, and so I periodically send an email asking the editor if any progress has been made, in an attempt to nudge them to nudge the publisher.

Of course, these are all small actions, and periodically feelings of futility and even despair overwhelm me. Yet I remind myself that despair is unproductive and continue on my quest to redefine my role-ideals in light of a changing climate and a changing context for action. As a mother, friend, volunteer, and professor, I do my best to pursue my role-ideals and to find meaningful and enjoyable actions to fulfill those ideals—whether they be baking a vegan carrot cake to celebrate a birthday or writing advocacy letters for the Climate Action Now app. And, of course, trying to bring my family and friends along.

• • •

RESPONSIBILITY: Don'ts and Dos

1. Don't get too caught up in blaming. Be forward-looking in offering up solutions.
2. Do find actions that feel right for you. Consider your power, privilege, interest, and ability to work collectively with family, friends, and other volunteers—and just what you can find time for and enjoy.
3. Do push yourself to go further to address climate solutions in each of your roles—for example, as a parent, volunteer, activist, friend, worker, churchgoer, or media consumer.

AFTERWORD

In writing this book, I set out to explore the meaning of individual action in the face of massive climate disruption and fossil fuel industry intransigence. I knew that much of the responsibility for our current crisis belonged with the industry and its government collaborators, but I still felt individual action must somehow have value. I landed on the idea that individual climate-friendly food, transport, and energy behaviors would have more impact if we intentionally persuaded our friends and family to adopt them alongside us. But I was still troubled by how one might get involved in the political activism that is so desperately needed to bring about systemic change.

I didn't see a pathway for me, and perhaps thousands of other concerned climate citizens, to join in political action. I had read about the lawsuits against the big oil companies, and while they were critically important, joining them didn't seem within reach. But then, during the years I was writing this book, I began to volunteer with the nonprofits Elders Climate Action and the Climate Reality Project. All of a sudden, political action was no longer limited in my mind to those who were bringing lawsuits against the fossil fuel behemoths. By volunteering for a climate organization, political action was made simple. I could ask my representatives to support climate legislation simply by sending the letters or making the calls suggested in the organization's weekly newsletters. And I could join in meetings the group organized with congressional staffers. Knowing that thousands of other climate volunteers were taking similar actions offered a ray of hope that we were *in this together*, and that together, we might make a difference.

Once the Climate Action Now app came online, activism became even easier. Whenever I was waiting in line at the pharmacy or my doctor's office or killing time as my elderly mother-in-law cogitated her next Scrabble word or simply too tired to do anything else in the evening, I could simply click on the Climate Action Now app on my phone and dash off a few messages to my state senator, congressperson, or a climate-recalcitrant business executive. Just as a city organic-waste-collection program enables residents to change their food waste practices or GivingTuesday changes our donation practices, a new tool such as the Climate Action Now app enables activists to be more active and even allows activism to spread across a broad swath of the population who otherwise might not be climate advocates.

My climate activism has evolved in other ways since the Climate Action Now app came online. I joined the organization's "core actors" who create sign-on letters advocating climate legislation for app users. I specialize in food and farm policy and organize Zoom action parties where "partiers" send letters to Congress supporting climate provisions in the US Farm Bill, funding for local food donation nonprofits, or the alt-protein industry. This is just the latest way that my plant-rich eating, composting, and internet "slacktivism" have spilled over into advocacy.

I have come to believe that people should get involved in the struggle for climate justice in a way that works for them. I have also come to believe the conservation psychologists who wrote, "In general, limiting climate change requires interventions at multiple levels and time scales: technology change and policy change are necessary, but do not obviate the importance of individual and household behavior, especially where these have the potential to push forward such systemic change; likewise, individual responses to climate change are necessary but must be supported and enabled by policy and structural change. Moreover, behavioral, cultural, technological, economic, and policy changes interact: None can be fully assessed without considering the others."[1]

Admittedly, I don't always follow up on the weekly actions suggested by my local Climate Reality chapter. And I still sometimes wake up in the middle of the night with a pit in my stomach, anxious about a climate-changed world. But I remind myself that I need to calm down and try to get some sleep, if only to be rested well enough the next day to push the boundaries of my roles as an academic and a climate-concerned citizen, to push myself to do more emailing, create more sign-on letters, and organize more action parties. And looking forward, I know I will continue to learn from my Climate Action Now, Elders Climate Action, and Climate Reality colleagues about how I can more effectively reach people in power, including through protests and civil disobedience. I can then share my knowledge with my family and friends and ask them to join in climate action alongside me. Still to this day, the oil companies continue to shamelessly obfuscate their role in fomenting the global climate crisis and spreading its vast injustices. But the voices and actions of the everyday climate-concerned citizens are being heard and seen by government and industry. Together we are making our voices louder and our actions more bold.

Notes

INTRODUCTION

1. Christine Lepisto, "Ask Me about Plalking," *Treehugger*, October 11, 2018, https://www.treehugger.com/ask-me-about-plalking-4855223.

2. Robin Wall Kimmerer, *Braiding Sweetgrass: Indigenous Wisdom, Scientific Knowledge and the Teachings of Plants* (Minneapolis: Milkweed Editions, 2013), 328.

3. Michael Mann, *The New Climate War* (New York: PublicAffairs, 2021).

4. Michele Micheletti and Dietlind Stolle, "Consumer Strategies in Social Movements," in *The Oxford Handbook of Social Movements*, ed. Donatella Della Porta and Mario Diani (Oxford: Oxford University Press, 2015); Clive Barnett et al., *Globalizing Responsibility: The Political Rationalities of Ethical Consumption* (West Sussex, UK: Wiley-Blackwell, 2011).

5. Lissy Friedman et al. "Tobacco Industry Use of Personal Responsibility Rhetoric in Public Relations and Litigation: Disguising Freedom to Blame as Freedom of Choice," *American Journal of Public Health* 105 (2014): e1–e11, https://doi.org/10.2105/AJPH.2014.302226.

6. "Heartland Institute," DeSmog, accessed November 24, 2020, https://www.desmogblog.com/heartland-institute.

7. Jason Samenow, "Heartland Institute Launches Campaign Linking Terrorism, Murder, and Global Warming Belief," *Washington Post*, May 4, 2012, https://www.washingtonpost.com/blogs/capital-weather-gang/post/heartland-institute-launches-campaign-linking-terrorism-murder-and-global-warming-belief/2012/05/04/gIQAJJ3Q1T_blog.html#pagebreak; "Heartland Institute," DeSmog.

8. "Energy and the Environment Explained," US Energy Information Administration, 2021, https://www.eia.gov/energyexplained/energy-and-the-environment/where-greenhouse-gases-come-from.php.

9. Finis Dunaway, "The 'Crying Indian' Ad That Fooled the Environmental Movement," *Chicago Tribune*, November, 21, 2017, https://www.chicagotribune.com/opinion/commentary/ct-perspec-indian-crying-environment-ads-pollution-1123-20171113-story.html.

10. Paul Griffin, *The Carbon Majors Database: CDP Carbon Majors Report 2017* (London: CDP, 2017), https://climateaccountability.org/pdf/CarbonMajorsRpt2017%20Jul17.pdf; Richard Heede, *Carbon Majors: Update of Top Twenty Companies 1965–2017* (Snowmass, CO: Climate Accountability Institute, 2019), https://climateaccountability.org/pdf/CAI%20PressRelease%20Top20%20Oct19.pdf; Naomi Oreskes, "Playing Dumb on Climate Change" *New York Times*, January 3, 2015, http://www.nytimes.com/2015/01/04/opinion/sunday/playing-dumb-on-climate-change.html?hp&action=click&pgtype=Homepage&module=c-column-top-span-region®ion=c-column-top-span-region&WT.nav=c-column-top-span-region&_r=0; Naomi Oreskes and Erik M. Conway, *Merchants of Doubt: How a Handful of Scientists Obscured the Truth on Issues from Tobacco Smoke to Climate Change* (New York: Bloomsbury, 2011); Geoffrey Supran and Naomi Oreskes, "Rhetoric and Frame Analysis of Exxonmobil's Climate Change Communications," *One Earth*, May 13, 2021, https://doi.org/10.1016/j.oneear.2021.04.014; Amy Westervelt,

"Campaigns So Successful They've Landed in Court," November 17, 2017, in *Drilled: Critical Frequency*, podcast, https://podcasts.google.com/feed/aHR0cHM6Ly9mZWVkcy5tZWdhcGhvbmUuZm0vQ0ZRWTIyOTE3NzIzOTU/episode/MDlhNDcyZjNlODQyNDM1NjkxMTQ1Mjg5MjE5ZThlOGQ?sa=X&ved=0CAUQkfYCahgKEwig5IObt571AhUAAAAAHQAAAAAQ-RA&hl=en.

11. Mann, *New Climate War*; Supran and Oreskes, "Rhetoric and Frame Analysis."

12. Mann, *New Climate War*; Supran and Oreskes, "Rhetoric and Frame Analysis."

13. Thomas Marlow, Sean Miller, and J. Timmons Roberts, "Bots and Online Climate Discourses: Twitter Discourse on President Trump's Announcement of U.S. Withdrawal from the Paris Agreement," *Climate Policy* 21, no. 6 (2021): 765–77, https://doi.org/10.1080/14693062.2020.1870098; Mann, *New Climate War*.

14. Supran and Oreskes, "Rhetoric and Frame Analysis," 9.

15. John Cook et al., *America Misled: How the Fossil Fuel Industry Deliberately Misled Americans about Climate Change* (Fairfax, VA: George Mason University Center for Climate Change Communication, 2019), https://www.climatechangecommunication.org/america-misled/; Mann, *New Climate War*; Westervelt, "Campaigns So Successful"; Walter Sinnott-Armstrong, "It's Not My Fault," in *Perspectives on Climate Change: Science, Economics, Politics, Ethics*, ed. Walter Sinnott-Armstrong and Richard Howarth (New Milford, CT: Emerald Group, 2005), 285–307; Laura Sullivan, "How Big Oil Misled the Public into Believing Plastic Would Be Recycled," NPR, 2020, https://www.npr.org/2020/09/11/897692090/how-big-oil-misled-the-public-into-believing-plastic-would-be-recycled; Supran and Oreskes, "Rhetoric and Frame Analysis."

16. Mann, *New Climate War*, 62.

17. Mann, *New Climate War*, 61.

18. Sinnott-Armstrong, "It's Not My Fault," 304.

19. "Cipher Newsmakers: Interview with U.S. Energy Secretary Jennifer Granholm," accessed September 29, 2021, https://www.youtube.com/watch?v=a758TwPYqT4.

20. Mann, *New Climate War*.

21. Kimmerer, *Braiding Sweetgrass*.

22. Annie Lowrey, "All That Performative Environmentalism Adds Up," *The Atlantic*, August 31, 2020, https://www.theatlantic.com/ideas/archive/2020/08/your-tote-bag-can-make-difference/615817/.

23. James Andreoni, "Impure Altruism and Donations to Public Goods: A Theory of Warm-Glow Giving," *Economic Journal* 100, no. 401 (1990): 464–77, https://doi.org/10.2307/2234133; Ros McLellan and Susan Steward, "Measuring Children and Young People's Wellbeing in the School Context," *Cambridge Journal of Education* 45, no. 3 (2015): 307–32, https://doi.org/10.1080/0305764X.2014.889659; Roland Menges, Carsten Schroeder, and Stefan Traub, "Altruism, Warm Glow and the Willingness-to-Donate for Green Electricity: An Artefactual Field Experiment," *Environmental and Resource Economics* 31, no. 4 (2005): 431–58, https://doi.org/10.1007/s10640-005-3365-y.

24. Kate Soper, "Re-thinking the 'Good Life': The Citizenship Dimension of Consumer Disaffection with Consumerism," *Journal of Consumer Culture* 7, no. 2 (2007): 205–29, https://doi.org/10.1177/1469540507077681.

25. Toshimasa Sone et al., "Sense of Life Worth Living (*Ikigai*) and Mortality in Japan: Ohsaki Study," *Psychosomatic Medicine* 70, no. 6 (2008): 709–15, https://doi.org/10.1097/PSY.0b013e31817e7e64; Marta Zaraska, "Boosting Our Sense of Meaning in Life Is an Often Overlooked Longevity Ingredient," *Washington Post*, January 3, 2021, https://www.washingtonpost.com/health/boosting-our-sense-of-meaning-in-life-is-an-often-overlooked-longevity-ingredient/2020/12/31/84871d32–29d4–11eb-8fa2–06e7cbb145c0_story.html.

26. Elizabeth Cripps, *Climate Change and the Moral Agent: Individual Duties in an Interdependent World* (Oxford: Oxford University Press, 2013): 166.

27. Peter Kalmus, *Being the Change: How to Live Well and Spark a Climate Revolution* (Gabriola Island, BC: New Society, 2017), https://peterkalmus.net/books/read-by-chapter-being-the-change/.

28. Simon Hattenstone, "The Transformation of Greta Thunberg," *The Guardian*, September 25, 2021, https://www.theguardian.com/environment/ng-interactive/2021/sep/25/greta-thunberg-i-really-see-the-value-of-friendship-apart-from-the-climate-almost-nothing-else-matters.

29. Maria Ojala, "Coping with Climate Change among Adolescents: Implications for Subjective Well-Being and Environmental Engagement," *Sustainability* 5, no. 5 (2013): 2191, http://www.mdpi.com/2071-1050/5/5/2191; Maria Ojala, "Hope and Climate Change: The Importance of Hope for Environmental Engagement among Young People," *Environmental Education Research* 18, no. 5 (2012): 625–42, http://dx.doi.org/10.1080/13504622.2011.637157; Maria Ojala, "How Do Children Cope with Global Climate Change? Coping Strategies, Engagement, and Well-Being," *Journal of Environmental Psychology* 32, no. 3 (2012): 225–33, https://doi.org/10.1016/j.jenvp.2012.02.004.

30. Stuart Capstick and Radhika Khosla, "Bridging the Gap: The Role of Equitable Low-Carbon Lifestyles," in *Emissions Gap Report 2020*, ed. (Nairobi: United Nations Environment Programme, 2020), 62–75; David Hagmann, Emily H. Ho, and George Loewenstein, "Nudging Out Support for a Carbon Tax," *Nature Climate Change* 9, no. 6 (2019): 484–89, https://doi.org/10.1038/s41558-019-0474-0; Alexander Maki et al., "Meta-Analysis of Pro-Environmental Behaviour Spillover," *Nature Sustainability* 2 (2019): 307–15, https://doi.org/10.1038/s41893-019-0263-9; Peter Newell, Freddie Dailey, and Michelle Twena, *Changing Our Ways: Behaviour Change and the Climate Crisis* (Cambridge: Cambridge Sustainability Commission on Scaling Behaviour Change, 2021), https://www.rapidtransition.org/resources/cambridge-sustainability-commission/; Seth Werfel, "Household Behaviour Crowds Out Support for Climate Change Policy When Sufficient Progress Is Perceived," *Nature Climate Change* 7, no. 7 (2017): 512–15, https://doi.org/10.1038/nclimate3316.

31. Amanda R. Carrico et al., "Putting Your Money Where Your Mouth Is: An Experimental Test of Pro-Environmental Spillover from Reducing Meat Consumption to Monetary Donations," *Environment and Behavior* 50, no. 7 (2017): 723–48, https://doi.org/10.1177/0013916517713067; Maki et al., "Meta-Analysis of Pro-Environmental Behaviour Spillover"; Nick Nash et al., "Climate-Relevant Behavioral Spillover and the Potential Contribution of Social Practice Theory," *Wiley Interdisciplinary Reviews: Climate Change* 8, no. 6 (2017): e481, https://doi.org/10.1002/wcc.481; Nick Nash et al., "Reflecting on Behavioral Spillover in Context: How Do Behavioral Motivations and Awareness Catalyze Other Environmentally Responsible Actions in Brazil, China, and Denmark?," *Frontiers in Psychology* 10, no. 788 (2019), https://doi.org/10.3389/fpsyg.2019.00788; Newell et al., *Changing Our Ways*; Andreas Nilsson, Magnus Bergquist, and P. Wesley Schultz, "Spillover Effects in Environmental Behaviors, across Time and Context: A Review and Research Agenda," *Environmental Education Research* 23, no. 4 (2017): 573–89, https://doi.org/10.1080/13504622.2016.1250148; Gregg Sparkman, Shahzeen Attari, and Elke Weber, "Moderating Spillover: Focusing on Personal Sustainable Behavior Rarely Hinders and Can Boost Climate Policy Support," *Energy Research and Social Science* 78 (August 2021): 1–9, https://doi.org/10.1016/j.erss.2021.102150.

32. Sebastian Levi, "Why Hate Carbon Taxes? Machine Learning Evidence on the Roles of Personal Responsibility, Trust, Revenue Recycling, and Other Factors across 23 European Countries," *Energy Research and Social Science* 73 (March 2021): 1–13, https://doi.org/10.1016/j.erss.2020.101883.

33. Sparkman, "Moderating Spillover."

34. Gordon Kraft-Todd et al., "Credibility-Enhancing Displays Promote the Provision of Non-Normative Public Goods," *Nature*, no. 563 (2018): 245–48, https://doi.org/10.1038/s41586-018-0647-4.

35. Sparkman, "Moderating Spillover"; Gregg Sparkman and Shahzeen Attari, "Credibility, Communication, and Climate Change: How Lifestyle Inconsistency and Do-Gooder Derogation Impact Decarbonization Advocacy," *Energy Research and Social Science* 59 (January 2020): 1–7, https://doi.org/10.1016/j.erss.2019.101290.

36. Shahzeen Attari, David H. Krantz, and Elke U. Weber, "Climate Change Communicators' Carbon Footprints Affect Their Audience's Policy Support," *Climatic Change* 154, no. 3 (2019): 529–45, https://doi.org/10.1007/s10584-019-02463-0.

37. Lowrey, "All That Performative Environmentalism."

38. David Marchese, "Greta Thunberg Hears Your Excuses. She Is Not Impressed," *New York Times*, October 30, 2020, https://www.nytimes.com/interactive/2020/11/02/magazine/greta-thunberg-interview.html?action=click&module=Editors%20Picks&pgtype=Homepage.

39. Bryan Bollinger and Kenneth Gillingham, "Peer Effects in the Diffusion of Solar Photovoltaic Panels," *Marketing Science* 31, no. 6 (2012): 900–12, https://doi.org/10.1287/mksc.1120.0727; Damon Centola, *Change: How to Make Big Things Happen* (New York: Little, Brown Spark, 2021); Damon Centola, *How Behavior Spreads: The Science of Complex Contagions* (Princeton, NJ: Princeton University Press, 2018); Robert H. Frank, *Under the Influence: Putting Peer Pressure to Work* (Princeton, NJ: Princeton University Press, 2020).

40. Claire Brouwer et al., "Communication Strategies for Moral Rebels: How to Talk about Change in Order to Inspire Self-Efficacy in Others," *WIREs Climate Change* (2002), https://doi.org/10.1002/wcc.781.

41. Leor Hackel and Gregg Sparkman, "Reducing Your Carbon Footprint Still Matters," Slate, October 26, 2018, https://slate.com/technology/2018/10/carbon-footprint-climate-change-personal-action-collective-action.html.

42. Hackel and Sparkman, "Reducing Your Carbon Footprint"; Newell et al., *Changing Our Ways*; Kristian Nielsen et al., "How Psychology Can Help Limit Climate Change," *American Psychologist* 76, no. 1 (2021), https://doi.org/10.1037/amp0000624.

43. Dale Jamieson, "Responsibility and Climate Change," *Global Justice* 8, no. 2 (2015): 23–42, https://www.theglobaljusticenetwork.org/index.php/gjn/article/view/86; Nielsen et al., "How Psychology Can Help."

44. Katherine Stroebe, Tom Postmes, and Carla A. Roos, "Where Did Inaction Go? Towards a Broader and More Refined Perspective on Collective Actions," *British Journal of Social Psychology* 58, no. 3 (2019): 649–67, https://doi.org/10.1111/bjso.12295.

45. Jamieson, "Responsibility and Climate Change," 41.

46. Erinn Gilson, "Vote with Your Fork? Responsibility for Food Justice," *Social Philosophy Today* 30 (2014): 113–30, https://doi.org/10.5840/socphiltoday20144215.

47. Liz Hamel et al., *The Kaiser Family Foundation / Washington Post Climate Change Survey* (San Francisco: Kaiser Family Foundation, 2019), https://www.kff.org/report-section/the-kaiser-family-foundation-washington-post-climate-change-survey-main-findings/.

48. Ilona M. Otto et al., "Social Tipping Dynamics for Stabilizing Earth's Climate by 2050," *Proceedings of the National Academy of Sciences* 117, no. 5 (2020): 2354–65, https://doi.org/10.1073/pnas.1900577117.

49. Ricarda Winkelmann et al., "Social Tipping Processes towards Climate Action: A Conceptual Framework," *Ecological Economics* 192 (February 2022): 1–14, https://doi.org/10.1016/j.ecolecon.2021.107242.

50. Karine Nyborg et al., "Social Norms as Solutions," *Science* 354, no. 6308 (2016): 42–43, https://doi.org/10.1126/science.aaf8317; Otto et al., "Social Tipping Dynamics."

51. "The Project," Count Us In, accessed October 15, 2020, https://www.count-us-in.org/en-b/project/.

52. Cripps, *Climate Change and the Moral Agent.*

53. Barnett et al., *Globalizing Responsibility.*

54. Jamieson, "Responsibility and Climate Change," 181.

55. Bob Edwards, John D. McCarthy, and Dane R. Mataic, "The Resource Context of Social Movements," in *The Wiley Blackwell Companion to Social Movements*, ed. David A. Snow et al. (Hoboken, NJ: Wiley, 2018), 79–97.

56. Cf. Megan E. Brooker and David S. Meyer, "Coalitions and the Organization of Collective Action," in *The Wiley Blackwell Companion to Social Movements*, ed. David A. Snow et al. (Hoboken, NJ: Wiley, 2018), 252–68; Dan Wang and Sarah Soule, "Social Movement Organizational Collaboration: Networks of Learning and the Diffusion of Protest Tactics, 1960–1995," *American Journal of Sociology* 117, no. 6 (2012): 1674–1722, https://www.journals.uchicago.edu/doi/abs/10.1086/664685.

57. Mann, *New Climate War.*

58. Diana Ivanova et al., "Environmental Impact Assessment of Household Consumption," *Journal of Industrial Ecology* 20, no. 3 (2016): 526–36, https://doi.org/10.1111/jiec.12371.

59. Newell et al., *Changing Our Ways.*

60. Newell et al., *Changing Our Ways*, italics in original.

61. "Solutions," Project Drawdown, accessed July 3, 2018, https://www.drawdown.org/solutions; "Paul Hawken," accessed January 29, 2022, https://paulhawken.com/; William Shurtleff and Akiko Aoyagi, *History of Erewhon: Natural Foods Pioneer in the United States (1966–2011)* (Lafayette, CA: Soyinfo Center, 2011).

62. Amanda Ravenhill, "'Drawdown' Starts at Presidio Graduate School," Presidio Graduate School, November 14, 2017, https://www.presidio.edu/blog/amanda-ravenhill.

63. "Solutions," Project Drawdown.

64. Mann, *New Climate War*; Supran and Oreskes, "Rhetoric and Frame Analysis."

65. Jennifer Kent, "Individualized Responsibility: 'If Climate Protection Becomes Everyone's Responsibility, Does It End up Being No-One's?,'" *Cosmopolitan Civil Societies* 1, no. 3 (2009), https://doi.org/10.5130/ccs.v1i3.1081.

66. Auden Schendler, "Worrying about Your Carbon Footprint Is Exactly What Big Oil Wants You to Do," *New York Times*, August 31, 2021, https://www.nytimes.com/2021/08/31/opinion/climate-change-carbon-neutral.html; "Nebraska Takes Aim at Colorado's Meat-Free Day by Declaring Pro-Meat Day," *The Guardian*, March 15, 2021, https://www.theguardian.com/us-news/2021/mar/15/nebraska-colorado-meatless-day-beef-industry.

67. Nielsen et al., "How Psychology Can Help."

68. "Solutions," Project Drawdown.

69. Monica Bahati Kuumba, "Population Policy in the Era of Globalisation: A Case of Reproductive Imperialism," *Agenda: Empowering Women for Gender Equity*, no. 48 (2001): 22–30, https://doi.org/10.2307/4066510.

70. Nielsen et al., "How Psychology Can Help."

71. Dana Fisher and Sohana Nasrin, "Climate Activism and Its Effects," *WIREs Climate Change* 12, no. 1 (2021): e683, https://doi.org/10.1002/wcc.683.

72. Fisher and Nasrin, "Climate Activism and Its Effects."

73. Nielsen et al., "How Psychology Can Help."

74. Joseph Poore and Thomas Nemecek, "Reducing Food's Environmental Impacts through Producers and Consumers," *Science* 360, no. 6392 (2018): 987–92, https://doi.org/10.1126/science.aaq0216.

75. Diana Ivanova et al., "Mapping the Carbon Footprint of E.U. Regions," *Environmental Research Letters* 12, no. 5 (2017): 054013, https://doi.org/10.1088/1748-9326/aa6da9;

"Sources of Greenhouse Gas Emissions," EPA website, accessed June 2, 2022, https://www.epa.gov/ghgemissions/sources-greenhouse-gas-emissions.

76. Capstick and Khosla, "Bridging the Gap;" Newell et al., *Changing Our Ways*; Alicia P. Q. Wittmeyer, "I Admire Vegetarians. It's a Choice I Won't Ever Make," *New York Times*, February, 15, 2020, https://www.nytimes.com/2020/02/15/opinion/sunday/vegetarian-vegan-meat.html?action=click&module=Opinion&pgtype=Homepage.

77. Nielsen et al., "How Psychology Can Help."

78. Ryan McCrimmon, "The Red Meat Issue Biden Won't Touch," *Politico*, May 23, 2021, https://www.politico.com/news/2021/05/23/biden-farmers-beef-climate-emissions-490237; "Nebraska Takes Aim."

79. Joni Ernst, "In the Left's War on Meat, I'll Stand up for Iowa Farmers," Joni Ernst website, April 16, 2021, https://www.ernst.senate.gov/public/index.cfm/2021/4/joni-ernst-in-the-left-s-war-on-meat-i-ll-stand-up-for-iowa-farmers.

80. Marco Springmann et al., "The Healthiness and Sustainability of National and Global Food Based Dietary Guidelines: Modelling Study," *BMJ* 370 (2020): m2322, https://doi.org/10.1136/bmj.m2322.

81. Michael Sainato, "'Coal Is Over': The Miners Rooting for the Green New Deal," *The Guardian*, August, 12, 2019, https://www.theguardian.com/environment/2019/aug/12/west-virginia-appalachia-miners-green-new-deal.

82. Climate Justice Alliance website, accessed June 8, 2021, https://climatejusticealliance.org/.

83. Elizabeth Cripps, "Population Ethics for an Imperfect World: Basic Justice, Reasonable Disagreement, and Unavoidable Value Judgements," *Journal of Development Studies* 57, no. 9 (2021): 1470–82, https://doi.org/10.1080/00220388.2021.1915478.

84. Seth Wynes and Kimberly A. Nicholas, "The Climate Mitigation Gap: Education and Government Recommendations Miss the Most Effective Individual Actions," *Environmental Research Letters* 12 (2017): 1–2, https://doi.org/10.1088/1748-9326/aa7541.

85. Monica Crippa et al., *Fossil $C0_2$ Emissions of All World Countries: 2020 Report* (Luxembourg: Publications Office of the European Union, 2020), https://doi.org/10.2760/143674.

86. Monica Bahati Kuumba, "Perpetuating Neo-Colonialism through Population Control: South Africa and the United States," *Africa Today* 40, no. 3 (1993): 79–85, http://www.jstor.org/stable/4186924; Kuumba, "Population Policy"; Jessica Pearce, "Mississippi Appendectomies: Reliving Our Pro-Eugenics Past," *Ms.*, October 28, 2020, https://msmagazine.com/2020/10/28/ice-immigration-mississippi-appendectomies-usa-eugenics-forced-coerced-sterilization/.

87. "Essential Tips for Talking about Project Drawdown's Health and Education Solution," Project Drawdown, accessed September 24, 2021, https://drawdown.org/news/insights/essential-tips-for-talking-about-project-drawdowns-health-and-education-solution.

88. Tim Gore, *Confronting Carbon Inequality* (Boston: Oxfam, 2020), https://www.oxfam.org/en/research/confronting-carbon-inequality.

89. Eric Godoy, "What's the Harm in Climate Change?," *Ethics, Policy and Environment* 20, no. 1 (2017): 103–17, https://doi.org/10.1080/21550085.2017.1291828.

90. Sparkman et al., "Moderating Spillover."

91. Capstick and Khosla, "Bridging the Gap"; Newell et al., *Changing Our Ways*.

92. Anju Aggarwal and Adam Drewnowski, "Plant- and Animal-Protein Diets in Relation to Sociodemographic Drivers, Quality, and Cost: Findings from the Seattle Obesity Study," *American Journal of Clinical Nutrition* 110, no. 2 (2019): 451–60, https://doi.org/10.1093/ajcn/nqz064.

93. Jonathan Meer and Benjamin Priday, *Generosity across the Income and Wealth Distribution* (Washington, DC: National Bureau of Economic Research, 2020), http://www.nber.org/papers/w27076.

94. Richard G. Wilkinson and Kate Pickett, *The Inner Level: How More Equal Societies Reduce Stress, Restore Sanity and Improve Everyone's Well-Being* (London: Penguin, 2018).

95. Degrowth website, accessed May 24, 2022, https://degrowth.info/degrowth; Wen Stephenson, "No Safe Options: A Conversation with Andreas Malm," *Los Angeles Review of Books*, January 5, 2021, https://lareviewofbooks.org/article/no-safe-options-a-conversation-with-andreas-malm/.

96. Centola, *How Behavior Spreads*; Frank, *Under the Influence.*

97. Bente Halkier, "A Practice Theoretical Perspective on Everyday Dealings with Environmental Challenges of Food Consumption," *Anthropology of Food*, September 2009, https://doi.org/10.4000/aof.6405; Andreas Reckwitz, "Toward a Theory of Social Practices: A Development in Culturalist Theorizing," *European Journal of Social Theory* 5, no. 2 (2002): 243–63, https://doi.org/10.1177/13684310222225432.

98. Manfred Milinski et al., "Stabilizing the Earth's Climate Is Not a Losing Game: Supporting Evidence from Public Goods Experiments," *Proceedings of the National Academy of Sciences* 103, no. 11 (2006): 3994–98, https://doi.org/10.1073/pnas.0504902103.

99. Albert Bandura, "Self-Efficacy: Toward a Unifying Theory of Behavioral Change," *Psychological Review* 84, no. 2 (1977): 191–215; Donatella della Porta and Mario Diani, *Social Movements: An Introduction* (Hoboken, NJ: John Wiley & Sons, 2020); James Kitts, "Mobilizing in Black Boxes: Social Networks and Participation in Social Movement Organizations," *Mobilization: An International Quarterly* 5, no. 2 (2000): 241–57, https://doi.org/10.17813/maiq.5.2.5408016w34215787.

100. Iris Marion Young, "Responsibility and Global Justice: A Social Connection Model," *Social Philosophy and Policy* 23, no. 1 (2006): 102–30, https://doi.org/10.1017/S0265052506060043; Iris Marion Young, *Responsibility for Justice* (Oxford: Oxford University Press, 2011); Robin Zheng, "What Is My Role in Changing the System? A New Model of Responsibility for Structural Injustice," *Ethical Theory and Moral Practice* 21, no. 4 (2018): 869–85, https://doi.org/10.1007/s10677-018-9892-8.

101. Marianne E. Krasny, ed., *Cornell Climate Online Fellows: Stories of Climate Engagement* (Ithaca, NY: Cornell University Civic Ecology Lab, 2019), https://5d887b1f-9a56-41e8-99ac-488199be6cfa.filesusr.com/ugd/1210bd_e86dddb7715c439db-c89a2c078bbe197.pdf.

102. T. K. Ahn and Elinor Ostrom, "Social Capital and Collective Action," in *Handbook of Social Capital*, ed. Dario Castiglione, Jan W. van Deth, and Guglielmo Wolleb (Oxford: Oxford University Press, 2008), 70–100; Elinor Ostrom, "Analyzing Collective Action," *International Association of Agricultural Economists* 41, no. 1 (2010): 155–66, https://doi.org/10.1111/j.1574-0862.2010.00497.x; Amy Poteete, Marco Janssen, and Elinor Ostrom, *Working Together: Collective Action, the Commons, and Multiple Methods in Practice* (Princeton, NJ: Princeton University Press, 2010).

1. EAT

1. Anna Herforth and Selena Ahmed, "The Food Environment, Its Effects on Dietary Consumption, and Potential for Measurement within Agriculture-Nutrition Interventions," *Food Security* 7, no. 3 (2015): 505–20, https://doi.org/10.1007/s12571-015-0455-8; Maartje P. Poelman and Ingrid H. M. Steenhuis, "Food Choices in Context," in *Context: The Effects of Environment on Product Design and Evaluation*, ed. Herbert L. Meiselman (Sawston, UK: Woodhead, 2019), 143–68; C. Pam Zhang, "Junk Food Was Our Love Language," *New York Times*, November 27, 2020, https://www.nytimes.com/2020/11/27/style/mod-

ern-love-junk-food-was-our-love-language.html?action=click&module=Editors%20 Picks&pgtype=Homepage.

2. "Solutions," Project Drawdown, accessed July 3, 2018, https://www.drawdown.org/solutions.

3. Walter Willett et al., "Food in the Anthropocene: The Eat; Lancet Commission on Healthy Diets from Sustainable Food Systems," *The Lancet* 393, no. 10170 (2019): 447–92, https://doi.org/10.1016/S0140-6736(18)31788-4; "Land Use in Agriculture by the Numbers," FAO, May 7, 2020, https://www.fao.org/sustainability/news/detail/en/c/1274219/.

4. Michael Clark et al., "Global Food System Emissions Could Preclude Achieving the 1.5° and 2°C Climate Change Targets," *Science* 370, no. 6517 (2020): 705–8, https://doi.org/10.1126/science.aba7357; Cheikh Mbow et al., "Food Security," in *Climate Change and Land: An IPCC Special Report on Climate Change, Desertification, Land Degradation, Sustainable Land Management, Food Security, and Greenhouse Gas Fluxes in Terrestrial Ecosystems*, ed. Priyadarshi Shukla et al. (n.p.: IPCC: 2019), https://www.ipcc.ch/srccl/chapter/chapter-5/.

5. Marco Springmann et al., "Analysis and Valuation of the Health and Climate Change Cobenefits of Dietary Change," *Proceedings of the National Academy of Sciences* 113, no. 15 (2016): 4146–51, https://doi.org/10.1073/pnas.1523119113.

6. "Measuring and Analyzing Greenhouse Gases: Behind the Scenes," National Oceanic and Atmospheric Administration, accessed June 23, 2021, https://gml.noaa.gov/outreach/behind_the_scenes/gases.html.

7. Mbow et al., "Food Security."

8. NRDC, *Less Beef, Less Carbon* (New York: NRDC, 2017), https://www.nrdc.org/sites/default/files/less-beef-less-carbon-ip.pdf.

9. "Greenhouse Gas Emissions," Environmental Protection Agency, May 16, 2022, https://www.epa.gov/ghgemissions/overview-greenhouse-gases.

10. "Number of Cattle Worldwide from 2012 to 2022 (in Million Head)," Statista, accessed September 14, 2021, https://www.statista.com/statistics/263979/global-cattle-population-since-1990/.

11. Martin Heller and Gregory Keoleian, "Greenhouse Gas Emission Estimates of U.S. Dietary Choices and Food Loss," *Journal of Industrial Ecology* 19, no. 3 (2015): 391–401, https://doi.org/10.1111/jiec.12174.

12. Stephen Clune, Enda Crossin, and Karli Verghese, "Systematic Review of Greenhouse Gas Emissions for Different Fresh Food Categories," *Journal of Cleaner Production* 140 (2017): 766–83, https://doi.org/10.1016/j.jclepro.2016.04.082; Carolyn Opio et al., *Greenhouse Gas Emissions from Ruminant Supply Chains: A Global Life Cycle Assessment* (Rome: Food and Agriculture Organization [FAO], 2013), http://www.fao.org/3/i3461e/i3461e.pdf.

13. Marco Springmann et al., "Options for Keeping the Food System within Environmental Limits," *Nature* 562, no. 7728 (2018): 519–25, https://doi.org/10.1038/s41586-018-0594-0; Marco Springmann et al., "The Healthiness and Sustainability of National and Global Food Based Dietary Guidelines: Modelling Study," *BMJ* 370 (2020): m2322, https://doi.org/10.1136/bmj.m2322; Willett et al., "Food in the Anthropocene."

14. David Tilman and Michael Clark, "Global Diets Link Environmental Sustainability and Human Health," *Nature* 515, no. 7528 (2014): 518–22, https://doi.org/10.1038/nature13959.

15. Mbow et al., "Food Security."

16. Erinn Gilson, "Vote with Your Fork? Responsibility for Food Justice," *Social Philosophy Today* 30 (2014): 113–30, https://doi.org/10.5840/socphiltoday20144215.

17. Springmann et al., "Healthiness and Sustainability."

18. Ashley Parker, "Fresh off Election Falsehoods, Republicans Serve up a Whopper about Biden," *Washington Post*, April 26, 2021, https://www.washingtonpost.com/politics/

biden-burger-falsehood/2021/04/26/9afca0be-a6a2-11eb-8d25-7b30e74923ea_story.html.

19. Alicia P. Q. Wittmeyer, "I Admire Vegetarians. It's a Choice I Won't Ever Make," *New York Times*, February 15, 2020, https://www.nytimes.com/2020/02/15/opinion/sunday/vegetarian-vegan-meat.html?action=click&module=Opinion&pgtype=Homepage.

20. Charles Levkoe, "The Food Movement in Canada: A Social Movement Network Perspective," *Journal of Peasant Studies* 41, no. 3 (2014): 385–403, https://doi.org/10.1080/03066150.2014.910766; Thomas Lyson, *Civic Agriculture: Reconnecting Farm, Food, and Community* (Lebanon, NH: Tufts University Press, 2004).

21. Carmen Sirianni and Lewis A. Friedland, *The Civic Renewal Movement: Community Building and Democracy in the United States* (Dayton, OH: Charles Kettering Foundation, 2005).

22. Daniel Miller, *A Theory of Shopping* (Ithaca, NY: Cornell University Press, 1998).

23. Alice Malpass et al., "Problematizing Choice: Responsible Consumers and Sceptical Citizens," in *Governance, Consumers and Citizens: Consumption and Public Life*, ed. Mark Bevir and Frank Trentmann (London: Palgrave Macmillan, 2007), 231–56.

24. Benjamin Gardner, Gert-Jan de Bruijn, and Phillippa Lally, "A Systematic Review and Meta-Analysis of Applications of the Self-Report Habit Index to Nutrition and Physical Activity Behaviours," *Annals of Behavioral Medicine* 42, no. 2 (2011): 174–87, https://doi.org/10.1007/s12160-011-9282-0; Jonathan van't Riet et al., "The Importance of Habits in Eating Behaviour: An Overview and Recommendations for Future Research," *Appetite* 57, no. 3 (2011): 585–96, https://doi.org/10.1016/j.appet.2011.07.010.

25. Bas Verplanken and Wendy Wood, "Interventions to Break and Create Consumer Habits," *Journal of Public Policy and Marketing* 25, no. 1 (2006): 90–103, https://doi.org/10.1509/jppm.25.1.90.

26. Phllippa Lally, A. Chipperfield, and Jane Wardle, "Healthy Habits: Efficacy of Simple Advice on Weight Control Based on a Habit-Formation Model," *International Journal of Obesity* 32, no. 4 (2008): 700–707, https://doi.org/10.1038/sj.ijo.0803771; Verplanken and Wood, "Interventions."

27. Bas Verplanken and Deborah Roy, "Empowering Interventions to Promote Sustainable Lifestyles: Testing the Habit Discontinuity Hypothesis in a Field Experiment," *Journal of Environmental Psychology* 45 (March 2016): 127–34, https://doi.org/10.1016/j.jenvp.2015.11.008.

28. Richard Thaler and Cass Sunstein, *Nudge: Improving Decisions about Health, Wealth, and Happiness* (New Haven, CT: Yale University Press, 2008).

29. Robert Cialdini, "Crafting Normative Messages to Protect the Environment," *Current Directions in Psychological Science* 12, no. 4 (2003): 105–9, https://doi.org/10.1111/1467-8721.01242; Gregg Sparkman and Gregory Walton, "Witnessing Change: Dynamic Norms Help Resolve Diverse Barriers to Personal Change," *Journal of Experimental Social Psychology* 82 (May 2019): 238–52, https://doi.org/10.1016/j.jesp.2019.01.007.

30. Cass Sunstein and Lucia Reisch, "Automatically Green: Behavioral Economics and Environmental Protection," *Harvard Environmental Law Review* 38, no. 1 (2014): 127–58, https://doi.org/10.2139/ssrn.2245657.

31. Riet et al., "Importance of Habits."

32. Verplanken and Wood, "Interventions."

33. Jess Eng, "Vegan Barbecue Is Carving Out a Place in Traditional Meat-Smoking Regions," *Washington Post*, September 2, 2021, https://www.washingtonpost.com/food/2021/09/02/vegan-barbecue-faux-meats/.

34. Gustavo Arellano, "Carne Asada, Hold the Meat: Why Latinos Are Embracing Vegan-Mexican Cuisine," NPR, July 10, 2018, https://www.npr.org/sections/

thesalt/2018/07/19/629629261/carne-asada-hold-the-meat-why-latinos-are-embracing-vegan-mexican-cuisine.

35. Justin McCarthy and Scott Dekoster, "Nearly One in Four in U.S. Have Cut Back on Eating Meat," January 27, 2020, https://news.gallup.com/poll/282779/nearly-one-four-cut-back-eating-meat.aspx; Cynthia Ogden et al., "Prevalence of Obesity among Adults, by Household Income and Education: United States, 2011–2014," Centers for Disease Control, *Morbidity and Mortality Weekly Report* 66, no. 50 (2017): 1369–73, http://dx.doi.org/10.15585/mmwr.mm6650a1.

36. George Reynolds, "Why Do People Hate Vegans?," *The Guardian*, October 15, 2019, https://www.theguardian.com/lifeandstyle/2019/oct/25/why-do-people-hate-vegans.

37. Mark Pachucki, Paul Jacques, and Nicholas Christakis, "Social Network Concordance in Food Choice among Spouses, Friends, and Siblings," *American Journal of Public Health* 101, no. 11 (2011): 2170–77, https://doi.org/10.2105/ajph.2011.300282.

38. Herforth and Ahmed, "Food Environment"; Poelman and Steenhuis, "Food Choices in Context."

39. Centola, *How Behavior Spreads.*

40. Mark Granovetter, "The Strength of Weak Ties," *American Journal of Sociology* 78, no. 6 (1974): 1360–80, https://www.jstor.org/stable/2776392.

41. Centola, *How Behavior Spreads.*

42. Centola, *How Behavior Spreads.*

43. Susan Clayton and Susan Opotow, *Identity and the Natural Environment* (Cambridge, MA: MIT Press, 2003).

44. Henri Tajfel and John C. Turner, "The Social Identity Theory of Intergroup Behavior," in *Psychology of Intergroup Relations*, ed. Stephen Worchel and William Austin (Chicago: Nelson Hall, 1986), 7–24.

45. Francesca Polletta and James M Jasper, "Collective Identity and Social Movements," *Annual Review of Sociology* 27 (2001): 283–305, http://www.jstor.org/stable/2678623; Clare Saunders, "Double-Edged Swords? Collective Identity and Solidarity in the Environment Movement," *British Journal of Sociology* 59, no. 2 (2008): 227–53, https://doi.org/10.1111/j.1468-4446.2008.00191.x.

46. Susan Clayton, "Environment and Identity," in *The Oxford Handbook of Environmental and Conservation Psychology*, ed. Susan Clayton (Oxford: Oxford University Press, 2012), 164–80.

47. Malpass et al., "Problematizing Choice."

48. Centola, *How Behavior Spreads.*

49. Matthew H. Goldberg, Abel Gustafson, and Sander van der Linden, "Leveraging Social Science to Generate Lasting Engagement with Climate Change Solutions," *One Earth* 3, no. 3 (2020): 314–24, https://doi.org/10.1016/j.oneear.2020.08.011.

50. Blanca Requero et al., "Promoting Healthy Eating Practices through Persuasion Processes," *Basic and Applied Social Psychology* 43, no. 4 (2021): 239–66, https://doi.org/10.1080/01973533.2021.1929987.

51. Robert H. Frank, *Under the Influence: Putting Peer Pressure to Work* (Princeton, NJ: Princeton University Press, 2020).

52. Frank, *Under the Influence*; Chad Mortensen et al., "Trending Norms: A Lever for Encouraging Behaviors Performed by the Minority," *Social Psychological and Personality Science* 10, no. 2 (2019): 201–10, https://doi.org/10.1177/1948550617734615; Gregg Sparkman and Gregory Walton, "Dynamic Norms Promote Sustainable Behavior, Even If It Is Counternormative," *Psychological Science* 28, no. 11 (2017): 1663–74, https://doi.org/10.1177/0956797617719950; Sparkman and Walton, "Witnessing Change."

53. Damon Centola, "The Truth about Behavioral Change," *MIT Sloan Management Review* 60, no. 2 (2018), https://sloanreview.mit.edu/article/the-truth-about-behavioral-change/.

54. *The Game Changers* website, accessed January 29, 2022, https://gamechangersmovie.com/.

55. "One Day a Week Eat Less Meat: Lyric Video," April 22, 2019, https://www.youtube.com/watch?v=QCpL8y__WUc.

56. Centola, *How Behaviors Spread*; Frank, *Under the Influence*.

57. Ilona M. Otto et al., "Social Tipping Dynamics for Stabilizing Earth's Climate by 2050," *Proceedings of the National Academy of Sciences* 117, no. 5 (2020): 2354–65, https://doi.org/10.1073/pnas.1900577117.

58. "10 Black Vegan Influencers Who Inspire Us Everyday," The Beet, June 5, 2020, https://thebeet.com/10-black-influencers-in-the-vegan-instagram-world-to-follow/.

59. David Remnick, "Alexandria Ocasio-Cortez Is Coming for Your Hamburgers," *New Yorker*, March 3, 2019, https://www.newyorker.com/news/daily-comment/alexandria-ocasio-cortez-is-coming-for-your-hamburgers.

60. Suzanne Higgs, "Social Norms and Their Influence on Eating Behaviours," *Appetite* 86 (March 2015): 38–44, https://doi.org/10.1016/j.appet.2014.10.021.

61. Higgs, "Social Norms."

62. Frank, *Under the Influence*; Diana Mutz and Lori Young, "Communication and Public Opinion: Plus Ça Change?," *Public Opinion Quarterly* 75, no. 5 (2011): 1018–44, https://doi.org/10.1093/poq/nfr052.

63. Nicholas Christakis and James Fowler, "The Spread of Obesity in a Large Social Network over 32 Years," *New England Journal of Medicine* 357 (July 26, 2007): 370–79, https://doi.org/10.1056/NEJMsa066082.

64. Christakis and Fowler, "Spread of Obesity"; Nicholas Christakis, and James Fowler, "Social Contagion Theory: Examining Dynamic Social Networks and Human Behavior," *Statistics in Medicine* 32, no. 4 (2013): 556–77, https://doi.org/10.1002/sim.5408; Eric Robinson et al., "What Everyone Else Is Eating: A Systematic Review and Meta-Analysis of the Effect of Informational Eating Norms on Eating Behavior," *Journal of the Academy of Nutrition and Dietetics* 114, no. 3 (2014): 414–29, https://doi.org/10.1016/j.jand.2013.11.009.

65. "Snapshot: Few Americans Vegetarian or Vegan," Gallup, August 1, 2018, https://news.gallup.com/poll/238328/snapshot-few-americans-vegetarian-vegan.aspx; "How Many Blacks, Latinos, and Asians Are Vegan and Vegetarian in the U.S.A.?" Vegetarian Resource Group, December 17, 2020, https://www.vrg.org/blog/2020/12/17/how-many-blacks-latinos-and-asians-are-vegan-and-vegetarian/.

66. "Exploring the Explosion of Veganism in the United States," Ipsos Retail Performance, accessed November 28, 2020, https://www.ipsos-retailperformance.com/en/vegan-trends/.

67. Robert Cialdini, Raymond Reno, and Carl Kallgren, "A Focus Theory of Normative Conduct: Recycling the Concept of Norms to Reduce Littering in Public Places," *Journal of Personality and Social Psychology* 58, no. 6 (1990): 1015–26, https://doi.org/10.1037/0022-3514.58.6.1015; Mortensen et al., "Trending Norms"; Jessica Nolan et al., "Normative Social Influence Is Underdetected," *Personality and Social Psychology Bulletin* 34, no. 7 (2008): 913–23, https://doi.org/doi:10.1177/0146167208316691; Sparkman, and Walton, "Witnessing Change."

68. Sparkman and Walton, "Dynamic Norms."

69. Everett Rogers, *Diffusion of Innovations*, 5th ed. (New York: Free Press, 2003).

70. Jason Carmichael and Robert Brulle, "Elite Cues, Media Coverage, and Public Concern: An Integrated Path Analysis of Public Opinion on Climate Change, 2001–2013," *Environmental Politics* 26, no. 2 (2017): 232–52, https://doi.org/10.1080/09644016.2016.1263433.

71. Mark Harvey, *Celebrity Influence: Politics, Persuasion, and Issue-Based Advocacy* (Lawrence: University Press of Kansas, 2018).

72. Chen Lou, "Social Media Influencers and Followers: Theorization of a Trans-Parasocial Relation and Explication of Its Implications for Influencer Advertising," *Journal of Advertising*, March 5, 2021, 1–18, https://doi.org/10.1080/00913367.2021.1880345; Paula Stehr et al., "Parasocial Opinion Leadership Media Personalities' Influence within Parasocial Relations: Theoretical Conceptualization and Preliminary Results," *International Journal of Communication* 9 (2015): 982–1001, https://ijoc.org/index.php/ijoc/article/view/2717.

73. Cassie Powney, "This Is Not a Drill, People," *Cosmopolitan*, January 21, 2020, https://www.cosmopolitan.com/uk/beauty-hair/celebrity-hair-makeup/a30606064/khloe-kardashian-nude-lipstick/.

74. Ethan Varian, "While California Fires Rage, the Rich Hire Private Firefighters," *New York Times*, October 26, 2019, https://www.nytimes.com/2019/10/26/style/private-firefighters-california.html.

75. "Ipsos Encyclopedia—Affluent," Ipsos, March 10, 2018, https://www.ipsos.com/en/ipsos-encyclopedia-affluent.

76. Thomas Wiedmann et al., "Scientists' Warning on Affluence," *Nature Communications* 11, no. 1 (2020): 3107, https://doi.org/10.1038/s41467-020-16941-y.

77. Frank, *Under the Influence.*

78. Tara Gallimore, "Understanding the Reasons for and Barriers to Becoming Vegetarian in Prospective Vegetarians and Vegans," (MS thesis, McGill University, 2016), https://escholarship.mcgill.ca/concern/theses/vd66w254b?locale=en; Susanne Stoll-Kleemann and Uta Johanna Schmidt, "Reducing Meat Consumption in Developed and Transition Countries to Counter Climate Change and Biodiversity Loss: A Review of Influence Factors," *Regional Environmental Change* 17, no. 5 (2017): 1261–77, https://doi.org/10.1007/s10113-016-1057-5.

79. Albert Bandura, "Self-Efficacy: Toward a Unifying Theory of Behavioral Change," *Psychological Review* 84, no. 2 (1977): 191–215; Marianne Krasny, *Advancing Environmental Education Practice* (Ithaca, NY: Cornell University Press, 2020), https://www.cornellpress.cornell.edu/book/9781501747076/advancing-environmental-education-practice/.

80. Jacqueline Frick, Florian G. Kaiser, and Mark Wilson, "Environmental Knowledge and Conservation Behavior: Exploring Prevalence and Structure in a Representative Sample," *Personality and Individual Differences* 37, no. 8 (2004): 1597–613, https://doi.org/10.1016/j.paid.2004.02.015.

81. Todd Rogers, Noah J. Goldstein, and Craig R. Fox, "Social Mobilization," *Annual Review of Psychology* 69, no. 1 (2018): 357–81, https://doi.org/10.1146/annurev-psych-122414-033718, 360–61.

82. Menus of Change, Culinary Institute of America, accessed June, 21, 2021, https://www.menusofchange.org/; "Plant-Rich Resources," Civic Ecology Lab, Cornell University, accessed December 23, 2021, https://www.civicecology.org/plant-rich-resources; Elizabeth L. Fox et al., "A Focused Ethnographic Study on the Role of Health and Sustainability in Food Choice Decisions," *Appetite* 165 (October 2021): 105319, https://doi.org/10.1016/j.appet.2021.105319; "Vegetarian Diet: How to Get the Best Nutrition," Mayo Clinic, accessed November 10, 2021, https://www.mayoclinic.org/healthy-lifestyle/nutrition-and-healthy-eating/in-depth/vegetarian-diet/art-20046446.

2. GLEAN

1. Alex V. Barnard, "'Waving the Banana' at Capitalism: Political Theater and Social Movement Strategy among New York's 'Freegan' Dumpster Divers," *Ethnography* 12 (November 2011): 419–44, https://doi.org/10.1177/1466138110392453, 420.

2. "Freeganism," Wikipedia, accessed June 5, 2021, https://en.wikipedia.org/wiki/Freeganism.

3. Barnard, "'Waving the Banana' at Capitalism."

4. Peter Alexander et al., "Losses, Inefficiencies and Waste in the Global Food System," *Agricultural Systems* 153 (May 2017): 190–200, https://doi.org/10.1016/j.agsy.2017.01.014; FAO, *Global Food Losses and Food Waste: Extent, Causes and Prevention* (Rome: FAO, 2011).

5. Brook Lyndhurst Ltd., *Household Waste Prevention Evidence Review* (London: Defra, 2009); Liz Goodwin and Mark Barthel, "Food for Thought," in *Sustainable Consumption: Stakeholder Perspectives* (Geneva: World Economic Forum, 2013), 33–41; Matt Watson and Angela Meah, "Food, Waste and Safety: Negotiating Conflicting Social Anxieties into the Practices of Domestic Provisioning," *Sociological Review* 60, no. 2 suppl. (2012): 102–20, https://doi.org/10.1111/1467-954X.12040.

6. ReFED, *A Roadmap to Reduce U.S. Food Waste by 20 Percent*, 2016, https://staging.refed.org/downloads/ReFED_Report_2016.pdf.

7. ReFED, *Roadmap to 2030: Reducing U.S. Food Waste by 50% and the Refed Insights Engine*, (n.d.), https://refed.org/uploads/refed_roadmap2030-FINAL.pdf.

8. Emilie Weiben, *Save Food for a Better Climate: Converting the Food Loss and Waste Challenge into Climate Action* (Rome: FAO, 2017), http://www.fao.org/3/i8000e/i8000e.pdf.

9. Cheikh Mbow et al., "Food Security," in *Climate Change and Land: An IPCC Special Report on Climate Change, Desertification, Land Degradation, Sustainable Land Management, Food Security, and Greenhouse Gas Fluxes in Terrestrial Ecosystems*, ed. Priyadarshi Shukla et al. (n.p.: IPCC, 2019), https://www.ipcc.ch/srccl/chapter/chapter-5/.

10. FAO, *The State of Food and Agriculture: Moving Forward on Food Loss and Waste Reduction* (Rome: FAO, 2019), https://www.fao.org/publications/card/en/c/ca6030en/.

11. Safoora Mirmohamadsadeghi et al., "Biogas Production from Food Wastes: A Review on Recent Developments and Future Perspectives," *Bioresource Technology Reports* 7 (September 2019): 100202, https://www.sciencedirect.com/science/article/abs/pii/S2589014X19300921?via%3Dihub

12. Pöyry Management Consulting Oy, *Food Waste to Biofuels* (Oslo: Nordic Energy Research, 2019), https://www.nordicenergy.org/publications/food-waste-to-biofuels/.

13. "Ecoeconnect," CASL Trust Zimbabwe, accessed June 28, 2021, https://casltrust.org/; Respect Musiyiwa, "Community-Based Advocacy: Actions at the Frontline," in *Cornell Climate Online Fellows: Stories of Climate Engagement*, ed. Marianne E. Krasny (Ithaca, NY: Cornell University Civic Ecology Lab, 2019), https://www.civicecology.org/publications, 44–48.

14. FAO, *Global Food Losses and Food Waste;* FAO, *State of Food and Agriculture*; Vivianne Visschers, Nadine Wickli, and Michael Siegrist, "Sorting Out Food Waste Behaviour: A Survey on the Motivators and Barriers of Self-Reported Amounts of Food Waste in Households," *Journal of Environmental Psychology* 45 (March 2016): 66–78, https://doi.org/10.1016/j.jenvp.2015.11.007.

15. Effie Papargyropoulou et al., "The Food Waste Hierarchy as a Framework for the Management of Food Surplus and Food Waste," *Journal of Cleaner Production* 76 (August 1, 2014): 106–15, https://doi.org/10.1016/j.jclepro.2014.04.020.

16. Papargyropoulou et al., "Food Waste Hierarchy."

17. Valeria A. Torok, Karen Luyckx, and Steven Lapidge, "Human Food Waste to Animal Feed: Opportunities and Challenges," *Animal Production Science* (June 8, 2021), https://doi.org/10.1071/AN20631.

18. Adapted from ReFED, *Roadmap to Reduce U.S. Food Waste.*

19. Anna Davies and David Evans, "Urban Food Sharing: Emerging Geographies of Production, Consumption and Exchange," *Geoforum* 99 (February 2019): 154–59, https://doi.org/10.1016/j.geoforum.2018.11.015; Johanna F. Gollnhofer, Henri A. Weijo, and John W. Schouten, "Consumer Movements and Value Regimes: Fighting Food Waste in

Germany by Building Alternative Object Pathways," *Journal of Consumer Research* 46, no. 3 (2019): 460–82, https://doi.org/10.1093/jcr/ucz004.

20. Jennifer Marston, "Surplus Food Marketplace Too Good to Go Raises $31M to Expand in the U.S.," The Spoon, January 7, 2021, https://thespoon.tech/surplus-food-marketplace-too-good-to-go-raises-31m-to-expand-in-the-u-s/; Adrienne Murray, "The Entrepreneur Stopping Food Waste," BBC News, January 6, 2020, https://www.bbc.com/news/business-50974009; Too Good To Go, accessed June 5, 2021, https://toogoodtogo.org/en/; Kate Yoder, "Food Waste Is Heating Up the Planet. Is Dumpster-Diving by App a Solution?" *Grist*, May 25, 2021, https://grist.org/food/food-waste-climate-change-too-good-to-go-dumpster-diving/.

21. Olio, accessed October 15, 2021, https://olioex.com/.

22. Papargyropoulou et al., "Food Waste Hierarchy."

23. Torok et al., "Human Food Waste to Animal Feed."

24. Papargyropoulou et al., "Food Waste Hierarchy."

25. Calthorpe Community Garden, accessed March 12, 2021, https://www.calthorpe-communitygarden.org.uk/environment.

26. European Biogas Association (EBA), *Biogas Success Stories 2020* (Brussels: EBA, 2020), https://www.europeanbiogas.eu/biogas-success-stories-2020/.

27. "Love Food Hate Waste," Waste and Resources Action Programme, accessed June 22, 2020, https://www.lovefoodhatewaste.com/.

28. Brook Lyndhurst Ltd., *Household Waste Prevention Evidence Review*; Tom Quested et al., "Spaghetti Soup: The Complex World of Food Waste Behaviours," *Resources, Conservation and Recycling* 79 (October 2013): 43–51, https://doi.org/10.1016/j.resconrec.2013.04.011; Daniel Welch, Joanne Swaffield, and David Evans, "Who's Responsible for Food Waste? Consumers, Retailers and the Food Waste Discourse Coalition in the United Kingdom," *Journal of Consumer Culture* 21, no. 2 (2018): 236–56, https://doi.org/10.1177/1469540518773801.

29. "Food Waste Monitor," ReFED, accessed October 19, 2021, https://insights-engine.refed.org/food-waste-monitor?break_by=sector&indicator=tons-surplus&view=detail&year=2016.

30. Thomas Madden, Pamela Scholder Ellen, and Icek Ajzen, "A Comparison of the Theory of Planned Behavior and the Theory of Reasoned Action," *Personality and Social Psychology Bulletin* 18, no. 1 (1992): 3–9, https://doi.org/10.1177/0146167292181001.

31. Icek Ajzen, "The Theory of Planned Behavior," *Organizational Behavior and Human Decision Processes* 50, no. 2 (1991): 179–211, https://doi.org/10.1016/0749-5978(91)90020-T; Madden et al., "Comparison of the Theory."

32. Paul van der Werf, Jamie A. Seabrook, and Jason A. Gilliland, "Food for Naught: Using the Theory of Planned Behaviour to Better Understand Household Food Wasting Behaviour," *Canadian Geographer / Le géographe canadien* 63, no. 3 (2019): 478–93, https://doi.org/10.1111/cag.12519.

33. Sally Russell et al., "Bringing Habits and Emotions into Food Waste Behaviour," *Resources, Conservation and Recycling* 125 (October 2017): 107–14, https://doi.org/10.1016/j.resconrec.2017.06.007.

34. Ella Graham-Rowe, Donna C. Jessop, and Paul Sparks, "Predicting Household Food Waste Reduction Using an Extended Theory of Planned Behaviour," *Resources, Conservation and Recycling* 101 (August 2015): 194–202, https://doi.org/10.1016/j.resconrec.2015.05.020; Christina M. Neubig et al., "Action-Related Information Trumps System Information: Influencing Consumers' Intention to Reduce Food Waste," *Journal of Cleaner Production* 261 (July 10, 2020): 121126, https://doi.org/10.1016/j.jclepro.2020.121126; Graham-Rowe et al., "Predicting Household Food Waste Reduction"; Russell et al., "Bringing Habits and Emotions into Food Waste Behaviour"; Karin Schanes, Karin Dobernig,

and Burcu Gözet, "Food Waste Matters: A Systematic Review of Household Food Waste Practices and Their Policy Implications," *Journal of Cleaner Production* 182 (May 1, 2018): 978–91, https://doi.org/10.1016/j.jclepro.2018.02.030; Violeta Stancu, Pernille Haugaard, and Liisa Lähteenmäki, "Determinants of Consumer Food Waste Behaviour: Two Routes to Food Waste," *Appetite* 96 (January 1, 2016): 7–17, https://doi.org/10.1016/j.appet.2015.08.025; Violeta Stefan et al., "Avoiding Food Waste by Romanian Consumers: The Importance of Planning and Shopping Routines," *Food Quality and Preference* 28, no. 1 (2013): 375–81, https://doi.org/10.1016/j.foodqual.2012.11.001; Paul van der Werf, Jamie Seabrook, and Jason Gilliland, "'Reduce Food Waste, Save Money': Testing a Novel Intervention to Reduce Household Food Waste," *Environment and Behavior* 53, no. 2 (2019): 151–83, https://doi.org/10.1177/0013916519875180; Visschers et al., "Sorting Out Food Waste Behaviour."

35. Stancu et al., "Determinants of Consumer Food Waste Behaviour"; van der Werf et al., "'Reduce Food Waste, Save Money'"; van der Werf et al., "Food for Naught"; Visschers et al., "Sorting Out Food Waste Behaviour."

36. Stefan et al., "Avoiding Food Waste."

37. Visschers et al., "Sorting Out Food Waste Behaviour."

38. Stancu et al., "Determinants of Consumer Food Waste Behaviour"; van der Werf et al., "'Reduce Food Waste, Save Money'"; van der Werf et al., "Food for Naught."

39. Graham-Rowe et al., "Predicting Household Food Waste Reduction."

40. Stefan et al., "Avoiding Food Waste."

41. Russell et al., "Bringing Habits and Emotions into Food Waste Behaviour."

42. Brook Lyndhurst Ltd., *Household Waste Prevention Evidence Review*; Ella Graham-Rowe, Donna C. Jessop, and Paul Sparks, "Identifying Motivations and Barriers to Minimising Household Food Waste," *Resources, Conservation and Recycling* 84 (March 2014): 15–23, https://doi.org/10.1016/j.resconrec.2013.12.005; Graham-Rowe et al, "Identifying Motivations and Barriers"; Schanes et al., "Food Waste Matters"; Watson and Meah, "Food, Waste and Safety."

43. Goodwin and Barthel, "Food for Thought."

44. Graham-Rowe et al., "Identifying Motivations and Barriers"; Stefan et al., "Avoiding Food Waste"; Visschers et al., "Sorting Out Food Waste Behaviour."

45. Bente Halkier, ed., *Consumption Challenged: Food in Medialised Everyday Lives* (Farnham, UK: Ashgate, 2010); Schanes et al., "Food Waste Matters."

46. Engineering National Academies of Sciences, and Medicine, *A National Strategy to Reduce Food Waste at the Consumer Level* (Washington, DC: National Academies Press, 2020), https://doi.org/10.17226/25876, 3.

47. Engineering National Academies, *National Strategy*.

48. Alan Metcalfe et al., "Food Waste Bins: Bridging Infrastructures and Practices," *Sociological Review* 60, no. 2 suppl. (2012): 135–55, https://doi.org/10.1111/1467-954X.12042.

49. Bente Halkier, "A Practice Theoretical Perspective on Everyday Dealings with Environmental Challenges of Food Consumption," *Anthropology of Food* (September 2009), https://doi.org/10.4000/aof.6405; Halkier, *Consumption Challenged*; Quested et al., "Spaghetti Soup"; Schanes et al., "Food Waste Matters."

50. Daniel Welch, "Behaviour Change and Theories of Practice: Contributions, Limitations and Developments," *Social Business* 7, nos. 3–4 (2016): 241–61, https://doi.org/10.1362/204440817X15108539431488.

51. Elizabeth Shove, Mika Pantzar, and Matt Watson, *The Dynamics of Social Practice: Everyday Life and How It Changes* (London: SAGE, 2012), https://doi.org/10.4135/9781446250655; Alan Warde, "Consumption and Theories of Practice," *Journal of Consumer Culture* 5, no. 2 (2005): 131–53, https://doi.org/10.1177/1469540505053090; Welch, "Social Practices and Behaviour Change."

52. Gert Spaargaren, "The Cultural Dimension of Sustainable Consumption Practices: An Exploration in Theory and Policy," in *Innovations in Sustainable Consumption: New Economics, Socio-Technical Transitions and Social Practices*, ed. Maurie J. Cohen, Halina Szejnwald Brown, and Philip J. Vergragt (Cheltenham, UK: Elgar, 2013), 229–51.

53. Sophie Dubuisson-Quellier and Séverine Gojard, "Why Are Food Practices Not (More) Environmentally Friendly in France? The Role of Collective Standards and Symbolic Boundaries in Food Practices," *Environmental Policy and Governance* 26, no. 2 (2016): 89–100, https://doi.org/10.1002/eet.1703; Halkier, "Practice Theoretical Perspective"; Halkier, *Consumption Challenged*; Hargreaves, "Practice-ing Behaviour Change: Applying Social Practice Theory to Pro-Environmental Behaviour Change," *Journal of Consumer Culture* 11 (March 25, 2011): 79–99, https://doi.org/10.1177/1469540510390500; Frank Trentmann, "Citizenship and Consumption," *Journal of Consumer Culture* 7, no. 2 (2007): 147–58, https://doi.org/10.1177/1469540507077667; Welch, "Social Practices and Behaviour Change."

54. "30 Incredible Recipes That Minimise Food Waste," NewsLifeMedia, accessed June 6, 2021, https://www.delicious.com.au/recipes/collections/gallery/30-incredible-recipes-that-minimise-food-waste/uxchdynj.

55. Dubuisson-Quellier and Gojard, "Why Are Food Practices"; Sarah Hards, "Social Practice and the Evolution of Personal Environmental Values," *Environmental Values* 20, no. 1 (2011): 23–42, https://doi.org/10.3197/096327111X12922350165996; Margit Keller, Bente Halkier, and Terhi-Anna Wilska, "Policy and Governance for Sustainable Consumption at the Crossroads of Theories and Concepts," *Environmental Policy and Governance* 26, no. 2 (2016): 75–88, https://doi.org/10.1002/eet.1702; Eric Klinenberg, Malcolm Araos, and Liz Koslov, "Sociology and the Climate Crisis," *Annual Review of Sociology* 46, no. 1 (2020): 649–69, https://doi.org/10.1146/annurev-soc-121919-054750; Theodore Schatzki, "Introduction: Practice Theory," in *Practice Turn in Contemporary Theory*, ed. Theodore Schatzki, Karin Knorr-Cetina, and Eike Savigny (Florence, KY: Routledge, 2000), 1–14; Welch, "Social Practices and Behaviour Change."

56. Elizabeth Shove and Mika Pantzar, "Consumers, Producers and Practices: Understanding the Invention and Reinvention of Nordic Walking," *Journal of Consumer Culture* 5, no. 1 (2005): 43–64, https://doi.org/10.1177/1469540505049846.

57. Kirsten Gram-Hanssen, "Understanding Change and Continuity in Residential Energy Consumption," *Journal of Consumer Culture* 11, no. 1 (2011): 61–78, https://doi.org/10.1177/1469540510391725; Marianne Krasny et al., "Civic Ecology Practices: Insights from Practice Theory," *Ecology and Society* 20, no. 2 (2015): 12, http://www.ecologyandsociety.org/vol20/iss2/art12/; Mika Pantzar and Elizabeth Shove, "Understanding Innovation in Practice: A Discussion of the Production and Reproduction of Nordic Walking," *Technology Analysis and Strategic Management* 22, no. 4 (2010): 447–61, https://doi.org/10.1080/09537321003714402; Andreas Reckwitz, "Toward a Theory of Social Practices: A Development in Culturalist Theorizing," *European Journal of Social Theory* 5, no. 2 (2002): 243–63; Warde, "Consumption and Theories of Practice."

58. "What to Do with Carrot Greens: 10 Inspiring Ideas," Oh My Veggies, accessed November 23, 2020, https://ohmyveggies.com/what-to-do-with-carrot-greens-inspiring-ideas/.

59. Gram-Hanssen, "Understanding Change and Continuity in Residential Energy Consumption"; Krasny et al., "Civic Ecology Practices"; Pantzar and Shove, "Understanding Innovation in Practice."

60. David Evans, Hugh Campbell, and Anne Murcott, "A Brief Pre-History of Food Waste and the Social Sciences," *Sociological Review* 60, no. S2 (2013): 5–26, https://doi.org/10.1111/1467-954X.12035 21; Graham-Rowe et al., "Predicting Household Food

Waste Reduction"; Quested et al., "Spaghetti Soup"; Stancu et al., "Determinants of Consumer Food Waste Behaviour"; Watson and Meah, "Food, Waste and Safety."

61. Stancu et al., "Determinants of Consumer Food Waste Behaviour"; Stefan et al., "Avoiding Food Waste."

62. Stancu et al., "Determinants of Consumer Food Waste Behaviour"; Stefan et al., "Avoiding Food Waste."

63. "Understanding Date Labels," FDA, accessed June 24, 2021, https://www.fdareader.com/blog/understanding-expiration-dates-and-date-labels.

64. Stancu et al., "Determinants of Consumer Food Waste Behaviour"; Visschers et al., "Sorting Out Food Waste Behaviour; van der Werf et al., "Food for Naught."

65. Keller et al., "Policy and Governance for Sustainable Consumption."

66. Marie Hebrok and Casper Boks, "Household Food Waste: Drivers and Potential Intervention Points for Design: An Extensive Review," *Journal of Cleaner Production* 151 (May 10, 2017): 380–92, https://doi.org/10.1016/j.jclepro.2017.03.069; Spaargaren, "Cultural Dimension"; Matt Watson et al., "Challenges and Opportunities for Re-Framing Resource Use Policy with Practice Theories: The Change Points Approach," *Global Environmental Change* 62 (May 2020): 102072, https://doi.org/10.1016/j.gloenvcha.2020.102072; Nick Nash et al., "Climate-Relevant Behavioral Spillover and the Potential Contribution of Social Practice Theory," *Wiley Interdisciplinary Reviews: Climate Change* 8, no. 6 (2017): e481, https://doi.org/10.1002/wcc.481.

67. David Evans, Daniel Welch, and Joanne Swaffield, "Constructing and Mobilizing 'the Consumer': Responsibility, Consumption and the Politics of Sustainability," *Environment and Planning A: Economy and Space* 49, no. 6 (2017): 1396–412, https://doi.org/10.1177/0308518X17694030; Love Food Hate Waste, Waste and Resources Action Programme, accessed June 22, 2020, https://www.lovefoodhatewaste.com/; Miranda J. Nicholes et al., "Surely You Don't Eat Parsnip Skins? Categorising the Edibility of Food Waste," *Resources, Conservation and Recycling* 147 (August 2019): 179–88, https://doi.org/10.1016/j.resconrec.2019.03.004; van der Werf et al., "'Reduce Food Waste, Save Money'"; "What to Do with Carrot Greens."

68. Spaargaren, "Cultural Dimension."

69. Rivka Galchen, "How South Korea Is Composting Its Way to Sustainability," *New Yorker*, March 2, 2020, https://www.newyorker.com/magazine/2020/03/09/how-south-korea-is-composting-its-way-to-sustainability.

70. Anna Bernstad, "Household Food Waste Separation Behavior and the Importance of Convenience," *Waste Management* 34, no. 7 (2014): 1317–23, https://doi.org/10.1016/j.wasman.2014.03.013; "Discover Denver Composts," City and County of Denver, accessed November 24, 2020, https://www.denvergov.org/content/denvergov/en/trash-and-recycling/composting/compost-collection-program.html; "Home Composting Services," Compost Now, accessed November 24, 2020, https://compostnow.org/home/.

71. Bernstad, "Household Food Waste Separation Behavior"; Pantzar and Shove, "Understanding Innovation in Practice."

72. Keller et al., "Policy and Governance for Sustainable Consumption."

73. Hajime Yamakawa et al., "Food Waste Prevention: Lessons from the Love Food, Hate Waste Campaign in the UK" (16th International Waste Management and Landfill Symposium, St. Margherita di Pula, Italy, October 2017), https://www.researchgate.net/publication/320331025_Food_waste_prevention_lessons_from_the_Love_Food_Hate_Waste_campaign_in_the_UK.

74. Laura Devaney and Anna R. Davies, "Disrupting Household Food Consumption through Experimental Homelabs: Outcomes, Connections, Contexts," *Journal of Consumer Culture* 17, no. 3 (2016): 823–44, https://doi.org/10.1177/1469540516631153; Nash et al., "Climate-Relevant Behavioral Spillover"; Quested et al., "Spaghetti Soup"; Schanes

et al., "Food Waste Matters"; Stancu et al., "Determinants of Consumer Food Waste Behaviour"; Stefan et al., "Avoiding Food Waste"; Watson and Meah, "Food, Waste and Safety."

75. "Borough Market," TripAdvisor, accessed July 10, 2020, https://www.tripadvisor.com/Attraction_Review-g186338-d260500-Reviews-Borough_Market-London_England.html.

76. Goodwin and Barthel, "Food for Thought."

77. Veronica Sharp, Sara Giorgi, and David Wilson, "Delivery and Impact of Household Waste Prevention Campaigns (at the Local Level)," *Waste Management and Research* 28, no. 3 (2010): 256–68, https://doi.org/10.1177/0734242X10361507.

78. "Love Food Hate Waste."

79. Brook Lyndhurst Ltd., *Household Waste Prevention Evidence Review*; Jayne Cox et al., "Household Waste Prevention: A Review of Evidence," *Waste Management and Research* 28, no. 3 (2010): 193–219, https://doi.org/10.1177/0734242X10361506.

80. Spaargaren, "Cultural Dimension."

81. "Save the Food," Natural Resources Defense Council, accessed July 10, 2020, https://savethefood.com/; StillTasty: Your Ultimate Shelf Life Guide, accessed July 10, 2020, https://www.stilltasty.com/.

82. Christian Reynolds et al., "Review: Consumption-Stage Food Waste Reduction Interventions: What Works and How to Design Better Interventions," *Food Policy* 83 (February 2019): 7–27, https://doi.org/10.1016/j.foodpol.2019.01.009; "Are You a Teacher Looking for Inspiration?" Waste and Resources Action Programme, accessed June 23, 2020, https://www.lovefoodhatewaste.com/article/are-you-teacher-looking-inspiration.

83. Goodwin and Barthel, "Food for Thought."

84. Quested et al., "Spaghetti Soup."

85. "City Harvest: Rescuing Food for New York's Hungry," City Harvest, accessed July 10, 2020, https://www.cityharvest.org/programs/food-rescue/.

86. "Food Rescue Locator," Sustainable America, accessed July 10, 2020, https://foodrescuelocator.com/.

87. National Gleaning Project, University of Vermont Law School, accessed July 10, 2020, https://nationalgleaningproject.org/.

88. Gollnhofer et al., "Consumer Movements and Value Regimes"; Leah Koenig, "The Classy Dive: The Dos and Don'ts of Dumpster Diving," in *The Real World Reader: A Rhetorical Reader for Writers*, ed. James S. Miller (Oxford: Oxford University Press, 2016), 74–77.

89. "Freegan Pony," Yelp, accessed December 9, 2021, https://www.yelp.com/biz/freegan-pony-paris-2.

90. "Freegan Pony: Will This Secret Restaurant Revolutionize the Paris Food Scene?," Secrets of Paris, accessed July 10, 2020, http://www.secretsofparis.com/heathers-secret-blog/freegan-pony-will-this-secret-restaurant-revolutionize-the-p.html.

91. "Freegan.Info: Strategies for Sustainable Living beyond Capitalism," Freegan.info, accessed July 10, 2020, https://freegan.info/what-is-a-freegan/freegan-philosophy/.

92. Gollnhofer et al., "Consumer Movements and Value Regimes."

93. Eric S. Godoy, "What's the Harm in Climate Change?," *Ethics, Policy and Environment* 20, no. 1 (2017): 112, https://doi.org/10.1080/21550085.2017.1291828.

94. Gollnhofer et al., "Consumer Movements and Value Regimes," 472.

95. Gollnhofer et al., "Consumer Movements and Value Regimes."

96. Tin Fy, "Taking Recycling One Step Further: China's Thriving Zero Waste Movement," *China Development Brief*, October 31, 2019, https://chinadevelopmentbrief.cn/reports/taking-recycling-one-step-further-how-the-zero-waste-movement-thrives-in-china/.

97. Donatella della Porta and Mario Diani, *Social Movements: An Introduction* (Hoboken, NJ: John Wiley & Sons, 2020); Erinn Gilson, "Vote with Your Fork? Responsibility for Food Justice," *Social Philosophy Today* 30 (2014): 113–30, https://doi.org/10.5840/socphiltoday20144215; Halkier, *Consumption Challenged*; Jennifer L. Wilkins, "Eating Right Here: Moving from Consumer to Food Citizen," *Agriculture and Human Values* 22, no. 3 (2005): 269–73, https://doi.org/10.1007/s10460-005-6042-4.

98. Evans et al., "Brief Pre-History of Food Waste"; Thomas Lyson, *Civic Agriculture: Reconnecting Farm, Food, and Community* (Lebanon, NH: Tufts University Press, 2004); Wilkins, "Eating Right Here."

99. Michele Micheletti, "Just Clothes? Discursive Political Consumerism and Political Participation," European Consortium for Political Research Joint Sessions, Workshop 24, Emerging Repertoires of Political Action: Toward a Systematic Study of Postcoventional Forms of Participation, Uppsala, Sweden, April 13–18, 2004).

100. Kaela Jubas, "Conceptual Con/Fusion in Democratic Societies: Understandings and Limitations of Consumer-Citizenship," *Journal of Consumer Culture* 7, no. 2 (2007): 231–54, https://doi.org/10.1177/1469540507077683; Halkier, *Consumption Challenged.*

101. Clive Barnett et al., *Globalizing Responsibility: The Political Rationalities of Ethical Consumption* (Chichester, UK: Wiley-Blackwell, 2011).

102. Barnett et al., *Globalizing Responsibility.*

103. Barnett et al., *Globalizing Responsibility.*

104. Keller et al., "Policy and Governance for Sustainable Consumption"; Spaargaren, "Cultural Dimension"; Gert Spaargaren, "Theories of Practices: Agency, Technology, and Culture; Exploring the Relevance of Practice Theories for the Governance of Sustainable Consumption Practices in the New World-Order," *Global Environmental Change* 21, no. 3 (2011): 813–22, https://doi.org/10.1016/j.gloenvcha.2011.03.010.

105. Barnett et al., *Globalizing Responsibility*; Micheletti, "Just Clothes?"

106. Anthony Giddens, *Modernity and Self-Identity: Self and Society in the Late Modern Age* (Stanford, CA: Stanford University Press, 1991).

107. Halkier, *Consumption Challenged.*

108. Chase DiBenedetto, "Greta Thunberg Asks for Less 'Blah, Blah, Blah' and More Honesty at COP26," Mashable, November 2, 2021, https://mashable.com/video/greta-thunberg-cop26-speech.

109. Celia Hatton, "Operation Empty Plate: China's Food Waste Campaigner," BBC, March 9, 2013, https://www.bbc.com/news/world-asia-china-21711928.

110. "China Launches 'Clean Plate' Campaign against Food Waste," BBC, August 13, 2020, https://www.bbc.com/news/world-asia-china-53761295?piano-modal.

111. "U.S. Food Waste Policy Finder," ReFED, accessed July1, 2021, https://policyfinder.refed.com/.

112. Johanna Gollnhofer and Daniel Boller, "The Evolution of the German Anti-Food Waste Movement: Turning Sustainable Ideas into Business," in *Food Waste Management: Solving the Wicked Problem*, ed. Elina Närvänen et al. (Cham, Switzerland: Palgrave Macmillan, 2020), 115–40.

113. Barnett et al., *Globalizing Responsibility.*

3. GIVE

1. Evan Comen, "The Size of a Home the Year You Were Born," *Wall Street Journal*, May 25, 2016, https://247wallst.com/special-report/2016/05/25/the-size-of-a-home-the-year-you-were-born/; US Census Bureau, *2015 Characteristics of New Housing* (Washington, DC: Department of Commerce, 2015); US Census Bureau, *Historical Households Tables* (Washington, DC: Department of Commerce, 2019), https://www.census.gov/data/tables/time-series/demo/families/households.html.

2. Robert H. Frank, *Under the Influence: Putting Peer Pressure to Work* (Princeton, NJ: Princeton University Press, 2020).

3. Kjell Arne Brekke and Olof Johansson-Stenman, "The Behavioural Economics of Climate Change," *Oxford Review of Economic Policy* 24, no. 2 (2008): 280–97, https://doi.org/10.1093/oxrep/grn012; Thomas Wiedmann et al., "Scientists' Warning on Affluence," *Nature Communications* 11, no. 1 (2020): 3107, https://doi.org/10.1038/s41467-020-16941-y.

4. "Ipsos Encyclopedia—Affluent," Ipsos, accessed November 23, 2020, https://www.ipsos.com/en/ipsos-encyclopedia-affluent.

5. Richard Wilkinson and Kate Pickett, *The Inner Level: How More Equal Societies Reduce Stress, Restore Sanity and Improve Everyone's Well-Being* (London: Penguin Books, 2018).

6. Andrew E. Clark, Paul Frijters, and Michael A. Shields, "Relative Income, Happiness, and Utility: An Explanation for the Easterlin Paradox and Other Puzzles," *Journal of Economic Literature* 46, no. 1 (2008): 95–144, https://doi.org/10.1257/jel.46.1.95; Robert H. Frank, "How Not to Buy Happiness," *Daedalus* 133, no. 2 (2004): 69–79, https://doi.org/10.1162/001152604323049415; Frank, *Under the Influence.*

7. Wilkinson and Pickett, *Inner Level.*

8. Elizabeth W. Dunn, Lara B. Aknin, and Michael I. Norton, "Prosocial Spending and Happiness: Using Money to Benefit Others Pays Off," *Current Directions in Psychological Science* 23, no. 1 (2014): 41–47, https://doi.org/10.1177/0963721413512503; John F. Helliwell et al., eds. *World Happiness Report 2020* (New York: Sustainable Development Solutions Network, 2020), https://worldhappiness.report/ed/2020/; Wilkinson and Pickett, *Inner Level.*

9. Erica Chenoweth, *Civil Resistance: What Everyone Needs to Know* (New York: Oxford University Press, 2021); Paul Engler, "Protest Movements Need the Funding They Deserve," *Stanford Social Innovation Review*, July 3, 2018, https://doi.org/10.48558/AYM1-SK19.

10. Clean Air Task Force, accessed January 2, 2022, https://www.catf.us/.

11. "THRIVE Agenda," Green New Deal Network, accessed November 23, 2020, https://www.thriveagenda.com/; Michael Sainato, "'Coal Is Over:': The Miners Rooting for the Green New Deal," *The Guardian*, August 12, 2019, https://www.theguardian.com/environment/2019/aug/12/west-virginia-appalachia-miners-green-new-deal.

12. Degrowth, accessed May 24, 2021, https://www.degrowth.info/en/.

13. Seth Wynes and Kimberly A. Nicholas, "The Climate Mitigation Gap: Education and Government Recommendations Miss the Most Effective Individual Actions," *Environmental Research Letters* 12, no. 7 (2017): 074024, https://doi.org/10.1088/1748-9326/aa7541.

14. "Solutions," Project Drawdown, accessed July 3, 2018, https://www.drawdown.org/solutions; Elina Pradhan, "Female Education and Childbearing: A Closer Look at the Data," World Bank, November 24, 2015, https://blogs.worldbank.org/health/female-education-and-childbearing-closer-look-data.

15. Monica Crippa et al., *Fossil CO_2 Emissions of All World Countries: 2020 Report* (Luxembourg: Publications Office of the European Union, 2020), https://edgar.jrc.ec.europa.eu/booklet/Fossil_CO2_emissions_of_all_world_countries_booklet_2020report.pdf.

16. Monica Bahati Kuumba, "Perpetuating Neo-Colonialism through Population Control: South Africa and the United States," *Africa Today* 40, no. 3 (1993): 79–85, http://www.jstor.org/stable/4186924.

17. Monica Bahati Kuumba, "Population Policy in the Era of Globalisation: A Case of Reproductive Imperialism," *Agenda: Empowering Women for Gender Equity*, no. 48 (2001): 22–30, https://doi.org/10.2307/4066510.

18. "Reproductive Justice," SisterSong Women of Color Reproductive Justice Collective, accessed October 28, 2020, https://www.sistersong.net/reproductive-justice/.

19. Saima S. Iqbal, "Louis Agassiz, under a Microscope," *Harvard Crimson*, March 18, 2021, https://www.thecrimson.com/article/2021/3/18/louis-agassiz-scrut/.

20. David Starr Jordan, *The Days of a Man: Volume 2* (Yonkers-on-Hudson, NY: World Book, 1922), 298.

21. "The Life, Works, and Eugenics Outlook of David Starr Jordan," accessed March 23, 2021, https://dsjeugenics.weebly.com/years-at-stanford-and-later-life.html; Matto Mildenberger, "The Tragedy of the *Tragedy of the Commons*," *Scientific American*, April 23, 2019, https://blogs.scientificamerican.com/voices/the-tragedy-of-the-tragedy-of-the-commons/; Lulu Miller, *Why Fish Don't Exist* (New York: Simon & Schuster, 2021); Jedediah Purdy, "Environmentalism's Racist History," *New Yorker*, August 13, 2015, https://www.newyorker.com/news/news-desk/environmentalisms-racist-history.

22. Femme International, accessed May 26, 2020, https://www.femmeinternational.org/.

23. "Family Planning/Contraception Methods," WHO, accessed May 24, 2021, https://www.who.int/news-room/fact-sheets/detail/family-planning-contraception.

24. "Solutions," Project Drawdown.

25. "Essential Tips for Talking about Project Drawdown's Health and Education Solution," Project Drawdown, accessed September 24, 2021, https://drawdown.org/news/insights/essential-tips-for-talking-about-project-drawdowns-health-and-education-solution; Christina Kwauk et al., *Girls' Education in Climate Strategies: Opportunities for Improved Policy and Enhanced Action in Nationally Determined Contributions* (Washington, DC: Brookings Institution, 2019); Malala Fund, *A Greener, Fairer Future: Why Leaders Need to Invest in Climate and Girls Education* (Washington, DC: Malala Fund, 2021), https://malala.org/newsroom/archive/malala-fund-publishes-report-on-climate-change-and-girls-education.

26. Wilkinson and Pickett, *Inner Level.*

27. "Essential Tips for Talking about Project Drawdown's Health and Education Solution."

28. Femme International.

29. Crippa et al., *Fossil CO_2 Emissions.*"

30. Barbara Tasch, "Ranked: The 30 Poorest Countries in the World," *Business Insider*, March 7, 2017, https://www.businessinsider.com/the-25-poorest-countries-in-the-world-2017-3#11-comoros-gdp-per-capita-1529-1247-20.

31. Greening Burundi Project, accessed May 25, 2021, https://www.greeningburundi.org/.

32. Team 54 Project, accessed May 26, 2021, https://www.team54project.org/.

33. "A Billion Begins with One," Nature Conservancy, accessed May 26, 2021, https://www.nature.org/en-us/get-involved/how-to-help/plant-a-billion/.

34. "Solutions," Project Drawdown.

35. Blanca Bernal, Lara T. Murray, and Timothy R. H. Pearson, "Global Carbon Dioxide Removal Rates from Forest Landscape Restoration Activities," *Carbon Balance and Management* 13, no. 1 (2018): 22, https://doi.org/10.1186/s13021-018-0110-8; Joseph E. Fargione et al., "Natural Climate Solutions for the United States," *Science Advances* 4, no. 11 (2018), https://doi.org/10.1126/sciadv.aat1869; Bronson W. Griscom et al., "Natural Climate Solutions," *Proceedings of the National Academy of Sciences* 114, no. 44 (2017): 11645–50, https://doi.org/10.1073/pnas.1710465114; Mark Sperow, "Carbon Sequestration Potential in Reclaimed Mine Sites in Seven East-Central States," *Journal of Environmental Quality* 35, no. 4 (2006): 1428–38; Songhan Wang et al., "Recent Global Decline

of CO2 Fertilization Effects on Vegetation Photosynthesis," *Science* 370, no. 6522 (2020): 1295–1300, https://doi.org/10.1126/science.abb7772.

36. Dianna Kopansky, "Peatlands Rewetting, Restoration and Conservation Offers a Low-Cost, Low-Tech, High Impact Nature-Based Solution for Climate Action," United Nations Environment Program, accessed June 1, 2022, https://wedocs.unep.org/bitstream/handle/20.500.11822/28893/Peatlands_Rewetting.pdf?sequence=1&isAllowed=y; Tim Searchinger et al., *Creating a Sustainable Food Future: A Menu of Solutions to Feed Nearly 10 Billion People by 2050* (Washington, DC: World Resources Institute, 2019), https://research.wri.org/sites/default/files/2019-07/WRR_Food_Full_Report_0.pdf.

37. Brady Dennis, "The Quest to Keep Carbon in North Carolina's Wetlands," *Washington Post*, May 31, 2022, https://www.washingtonpost.com/climate-solutions/2022/05/31/north-carolina-peat-carbon-capture/.

38. Hans Nicholas Jong, "Indonesia Forest-Clearing Ban Is Made Permanent, but Labeled 'Propaganda,'" Mongabay, August 14, 2019, https://news.mongabay.com/2019/08/indonesia-forest-clearing-ban-is-made-permanent-but-labeled-propaganda/.

39. "BBC News State of the Planet: Coalfield Development," BBC, December 2020, Coalfield Development, https://coalfield-development.org/videos/.

40. "Flood Risk Overview for West Virginia," First Street Foundation, accessed March 23, 2021, https://floodfactor.com/state/westvirginia/54_fsid.

41. Christopher Flavelle, "As Manchin Blocks Climate Plan, His State Can't Hold Back Floods," *New York Times*, October 17, 2021, https://www.nytimes.com/2021/10/17/climate/manchin-west-virginia-flooding.html.

42. Griscom et al., "Natural Climate Solutions"; Catherine V. Moore, "The Hopeful Work of Turning Appalachia's Mountaintop Coal Mines into Farms," *Yes!*, October 12, 2017, https://www.yesmagazine.org/issue/just-transition/2017/10/12/the-hopeful-work-of-turning-appalachias-mountaintop-coal-mines-into-farms.

43. Dario Kenner, *Carbon Inequality: The Role of the Richest in Climate Change* (New York: Routledge, 2019).

44. Elinor Ostrom, "Analyzing Collective Action," *International Association of Agricultural Economists* 41, no. s1 (2010): 155–66, https://onlinelibrary.wiley.com/doi/full/10.1111/j.1574-0862.2010.00497.x.

45. Garrett Hardin, "The Tragedy of the Commons," *Science* 162, no. 3859 (1968): 1243–48 https://doi.org/10.1126/science.162.3859.1243.

46. Hardin, "Tragedy of the Commons," 1246.

47. Mildenberger, "Tragedy of the *Tragedy of the Commons*."

48. Garrett Hardin, "Commentary: Living on a Lifeboat," *BioScience* 24, no. 10 (1974): 561–68, https://doi.org/10.2307/1296629.

49. Thomas Dietz, Elinor Ostrom, and Paul C. Stern, "The Struggle to Govern the Commons," *Science* 302, no. 5652 (2003): 1907–12, https://doi.org/10.1126/science.1091015.

50. "Garrett Hardin," Southern Poverty Law Center, accessed May 10, 2021, https://www.splcenter.org/fighting-hate/extremist-files/individual/garrett-hardin.

51. Manfred Milinski et al., "Stabilizing the Earth's Climate Is Not a Losing Game: Supporting Evidence from Public Goods Experiments," *Proceedings of the National Academy of Sciences* 103, no. 11 (2006): 3994–98, https://doi.org/10.1073/pnas.0504902103; Elinor Ostrom, *A Polycentric Approach for Coping with Climate Change* (Washington, DC: World Bank, 2009), https://openknowledge.worldbank.org/bitstream/handle/10986/4287/WPS5095.pdf; Amy R. Poteete, Marco A. Janssen, and Elinor Ostrom, *Working Together: Collective Action, the Commons, and Multiple Methods in Practice* (Princeton, NJ: Princeton University Press, 2010).

52. Poteete, Janssen, and Ostrom, *Working Together*; Sujoy Chakravarty et al., "4 Public Goods Experiments and Social Preferences," *Economic and Political Weekly* 46, no. 35 (2011): 61–70, https://www.jstor.org/stable/23017909.

53. Oliver P. Hauser et al., "Cooperating with the Future," *Nature* 511, no. 7508 (2014): 220–23, https://doi.org/10.1038/nature13530.

54. Manfred Milinski et al., "The Collective-Risk Social Dilemma and the Prevention of Simulated Dangerous Climate Change," *Proceedings of the National Academy of Sciences* 105, no. 7 (2008): 2291–94, https://doi.org/10.1073/pnas.0709546105.

55. Dale Jamieson, *Reason in a Dark Time* (Oxford: Oxford University Press, 2014); Dale Jamieson, "Responsibility and Climate Change," *Global Justice* 8, no. 2 (2015): 23–42, https://www.theglobaljusticenetwork.org/index.php/gjn/article/view/86; Milinski et al., "Collective-Risk Social Dilemma."

56. "Giving USA 2020," Giving USA, June 16, 2020, https://givingusa.org/giving-usa-2020-charitable-giving-showed-solid-growth-climbing-to-449-64-billion-in-2019-one-of-the-highest-years-for-giving-on-record/.

57. Helene Desanlis et al., *Funding Trends 2021: Climate Change Mitigation Philanthropy* (San Francisco: Climateworks Foundation, 2021), https://www.climateworks.org/wp-content/uploads/2021/10/CWF_Funding_Trends_2021.pdf.

58. Piia Lundberg et al., "Materialism, Awareness of Environmental Consequences and Environmental Philanthropic Behaviour among Potential Donors," *Environmental Values* 28, no. 6 (2019): 741–62, https://doi.org/10.3197/096327119X15579936382527.

59. "Giving USA 2020."

60. Adam Smith, *Theory of Moral Sentiments* (Glasgow: Strand & Edinburgh, 1759), http://metalibri.wikidot.com/title:theory-of-moral-sentiments:smith-a; "Adam Smith (1723–1790)," *Internet Encyclopedia of Philosophy*, accessed May 26, 2020, https://www.iep.utm.edu/smith/#H2; Brekke and Johansson-Stenman, "Behavioural Economics of Climate Change."

61. Anya Samak and Roman M. Sheremeta, *Recognizing Contributors: An Experiment on Public Goods*, ESI Working Paper, 13–34, http://digitalcommons.chapman.edu/esi_working_papers/26; Jason F. Shogren and Laura O. Taylor, "On Behavioral-Environmental Economics," *Review of Environmental Economics and Policy* 2, no. 1 (2008): 26–44, https://doi.org/10.1093/reep/rem027; Ganga Shreedhar and Susana Mourato, "Experimental Evidence on the Impact of Biodiversity Conservation Videos on Charitable Donations," *Ecological Economics* 158 (April 2019): 180–93, https://doi.org/10.1016/j.ecolecon.2019.01.001.

62. Beth Breeze, "How Donors Choose Charities: The Role of Personal Taste and Experiences in Giving Decisions," *Voluntary Sector Review* 4, no. 2 (2013): 165–83, http://dx.doi.org/10.1332/204080513X667792; Dunn et al., "Prosocial Spending and Happiness"; Milinski et al., "Stabilizing the Earth's Climate"; Samak and Sheremeta, *Recognizing Contributors*; Shogren, and Taylor, "On Behavioral-Environmental Economics"; Roger Tyers, "Nudging the Jetset to Offset: Voluntary Carbon Offsetting and the Limits to Nudging," *Journal of Sustainable Tourism* 26, no. 10 (2018): 1668–86, https://doi.org/10.1080/09669582.2018.1494737.

63. Tally Katz-Gerro et al., "Environmental Philanthropy and Environmental Behavior in Five Countries: Is There Convergence among Youth?," *Voluntas: International Journal of Voluntary and Nonprofit Organizations* 26, no. 4 (2015): 1485–1509, https://www.jstor.org/stable/43654663.

64. Xinfang Song et al., "A Survey of Game Theory as Applied to Social Networks," *Tsinghua Science and Technology* 25, no. 6 (2020): 734–42, https://doi.org/10.26599/TST.2020.9010005.

65. Damon Centola, *Change: How to Make Big Things Happen* (New York: Little, Brown Spark, 2021); Jeremy Heimans and Henry Timms, *New Power* (New York: Doubleday, 2018).

66. GivingTuesday, accessed December 19, 2021, https://www.givingtuesday.org/about/; Heimans and Timms, *New Power*.

67. Marco Tomassini and Alberto Antonioni, "Public Goods Games on Coevolving Social Network Models," *Frontiers in Physics* 8, no. 58 (2020), https://doi.org/10.3389/fphy.2020.00058.

68. Samak and Sheremeta, *Recognizing Contributors*.

69. Jack Krawczyk and Jon Steinberg, "How Content Is Really Shared: Close Friends, Not 'Influencers,'" *Advertising Age*, March 7, 2012, https://adage.com/article/digitalnext/content-shared-close-friends-influencers/233147.

70. Heimans and Timms, *New Power*, 43.

71. Milinski et al., "Stabilizing the Earth's Climate."

72. Michaël Aklin and Matto Mildenberger, "Prisoners of the Wrong Dilemma: Why Distributive Conflict, Not Collective Action, Characterizes the Politics of Climate Change," *Global Environmental Politics* 20, no. 4 (2020): 4–27, https://doi.org/10.1162/glep_a_00578; Robinson Meyer, "The Weekly Planet: An Outdated Idea Is Still Shaping Climate Policy," *The Atlantic*, April 20, 2021, https://www.theatlantic.com/science/archive/2021/04/an-outdated-idea-is-still-shaping-climate-policy/618652/.

73. Thomas Hale, "Catalytic Cooperation," *Global Environmental Politics* 20, no. 4 (2020): 73–98, https://doi.org/10.1162/glep_a_00561.

74. Leon Festinger, "Cognitive Dissonance," *Scientific American* 207, no. 4 (1962): 93–106, http://www.jstor.org/stable/24936719.

75. Brekke and Johansson-Stenman, "Behavioural Economics of Climate Change"; Timo Goeschl et al., "How Much Can We Learn about Voluntary Climate Action from Behavior in Public Goods Games?," *Ecological Economics* 171 (May 2020), https://doi.org/10.1016/j.ecolecon.2020.106591; Milinski et al., "Stabilizing the Earth's Climate."

76. Dean Karlan, John A. List, and Eldar Shafir, "Small Matches and Charitable Giving: Evidence from a Natural Field Experiment," *Journal of Public Economics* 95, no. 5 (2011): 344–50, https://doi.org/10.1016/j.jpubeco.2010.11.024.

77. Daniel Rondeau and John A. List, "Matching and Challenge Gifts to Charity: Evidence from Laboratory and Natural Field Experiments," *Experimental Economics* 11, no. 3 (2008): 253–67, https://doi.org/10.1007/s10683-007-9190-0.

78. Brekke and Johansson-Stenman, "Behavioural Economics of Climate Change."

79. Tanya Drollinger, "A Theoretical Examination of Giving and Volunteering Utilizing Resource Exchange Theory," *Journal of Nonprofit and Public Sector Marketing* 22, no. 1 (2010): 55–66, https://doi.org/10.1080/10495140903190416.

80. Milinski et al., "Stabilizing the Earth's Climate."

81. Jen Shang and Rachel Croson, "A Field Experiment in Charitable Contribution: The Impact of Social Information on the Voluntary Provision of Public Goods," *Economic Journal* 119, no. 540 (2009): 1422–39, https://doi.org/10.1111/j.1468-0297.2009.02267.x.

82. Francisco Alpizar, Fredrik Carlsson, and Olof Johansson-Stenman, "Anonymity, Reciprocity, and Conformity: Evidence from Voluntary Contributions to a National Park in Costa Rica," *Journal of Public Economics* 92, no. 5 (2008): 1047–60, https://doi.org/10.1016/j.jpubeco.2007.11.004.

83. Milinski et al., "Stabilizing the Earth's Climate."

84. Timo Goeschl et al., "From Social Information to Social Norms: Evidence from Two Experiments on Donation Behaviour," *Games* 9, no. 4 (2018): 91, https://www.mdpi.com/2073-4336/9/4/91; Marie Claire Villeval, "Public Goods, Norms and Cooperation,"

in *Handbook of Experimental Game Theory*, ed. C. Monica Capra et al. (Cheltenham, UK: Edward Elgar, 2020), 183–212.

85. Brekke and Johansson-Stenman, "Behavioural Economics of Climate Change."

86. Shang and Croson, "Field Experiment in Charitable Contribution."

87. Elinor Ostrom, "Collective Action and the Evolution of Social Norms," *Journal of Economic Perspectives* 14, no. 3 (2000): 137–58, https://doi.org/10.1257/jep.14.3.137.

88. Hauser et al., "Cooperating with the Future"; Tomassini and Antonioni, "Public Goods Games."

89. Hauser et al., "Cooperating with the Future."

90. Scott I. Rick, Beatriz Pereira, and Katherine A. Burson, "The Benefits of Retail Therapy: Making Purchase Decisions Reduces Residual Sadness," *Journal of Consumer Psychology* 24, no. 3 (2014): 373–80, https://doi.org/10.1016/j.jcps.2013.12.004.

91. James Andreoni, "Impure Altruism and Donations to Public Goods: A Theory of Warm-Glow Giving," *Economic Journal* 100, no. 401 (1990): 464–77, https://doi.org/10.2307/2234133; Soyoung Q. Park et al., "A Neural Link between Generosity and Happiness," *Nature Communications* 8, no. 1 (2017): 15964, https://doi.org/10.1038/ncomms15964.

92. Lundberg et al., "Materialism, Awareness"; Paul C. Stern and Thomas Dietz, "The Value Basis of Environmental Concern," *Journal of Social Issues* 50, no. 3 (1994): 65–84, https://doi.org/10.1111/j.1540-4560.1994.tb02420.x.

93. Gregory R. Maio et al., "Ideologies, Values, Attitudes, and Behavior," in *Handbook of Social Psychology*, ed. John Delamater (Boston: Springer, 2006), 283–308; Linda Steg et al., "General Antecedents of Personal Norms, Policy Acceptability, and Intentions: The Role of Values, Worldviews, and Environmental Concern," *Society and Natural Resources* 24, no. 4 (2011): 349–67, https://doi.org/10.1080/08941920903214116; Bas Verplanken and Rob W. Holland, "Motivated Decision Making: Effects of Activation and Self-Centrality of Values on Choices and Behavior," *Journal of Personality and Social Psychology* 82, no. 3 (2002): 434–47, https://doi.org/10.1037/0022-3514.82.3.434.

94. Andreoni, "Impure Altruism"; John W. Mayo and Catherine H. Tinsley, "Warm Glow and Charitable Giving: Why the Wealthy Do Not Give More to Charity?," *Journal of Economic Psychology* 30, no. 3 (2009): 490–99, https://doi.org/10.1016/j.joep.2008.06.001; Roland Menges, Carsten Schroeder, and Stefan Traub, "Altruism, Warm Glow and the Willingness-to-Donate for Green Electricity: An Artefactual Field Experiment," *Environmental and Resource Economics* 31, no. 4 (2005): 431–58, https://doi.org/10.1007/s10640-005-3365-y.

95. Elizabeth W. Dunn, Lara B. Aknin, and Michael I. Norton, "Spending Money on Others Promotes Happiness," *Science* 319 (2008): 1687–88, doi: 10.1126/science.1150952.

96. Dunn et al., "Prosocial Spending and Happiness."

97. Dunn et al., "Spending Money on Others," 1688.

98. James Andreoni, "Impure Altruism."

99. Park et al., "Neural Link."

100. Christina M. Karns, William E. Moore, and Ulrich Mayr, "The Cultivation of Pure Altruism via Gratitude: A Functional MRI Study of Change with Gratitude Practice," *Frontiers in Human Neuroscience* 11, no. 599 (2017), https://doi.org/10.3389/fnhum.2017.00599.

101. Kara Swisher, *Charlotte and Dave Willner, plus Airbnb CEO Brian Chesky*, podcast, July 9, 2018, Recode Decode, https://www.vox.com/2018/7/9/17542740/transcript-charlotte-dave-willner-facebook-immigration-airbnb-ceo-brian-chesky-recode-decode.

102. Erin Chan Ding, "'Crises Reveal': The Pandemic Changed How These Women Choose to Spend Their Money," *Washington Post*, October 21, 2021, https://www.washingtonpost.com/business/2021/10/21/money-values-in-the-pandemic/.

103. Anthony Leiserowitz et al., *Climate Change in the American Mind: April 2020* (New Haven, CT: Yale University, 2020), https://climatecommunication.yale.edu/publications/climate-change-in-the-american-mind-april-2020/toc/2/.

104. Desanlis et al., *Funding Trends 2021*.

4. VOLUNTEER

1. Climate Action Now, accessed January 14, 2022, https://climateactionnow.com/.

2. Eitan Hersh, *Politics Is for Power: How to Move beyond Political Hobbyism, Take Action, and Make Real Change* (New York: Scribner, 2020).

3. Mario Diani, "Social Movements and Collective Action," in *The SAGE Handbook of Social Network Analysis*, ed. J Scott and P. J. Carrington, 223–35 (London: SAGE, 2014); Scott C. Ganz and Sarah A. Soule, "Greening the Congressional Record: Environmental Social Movements and Expertise-Based Access to the Policy Process," *Environmental Politics* 28, no. 4 (2019): 685–706, https://doi.org/10.1080/09644016.2019.1565463.

4. Donatella della Porta and Mario Diani, *Social Movements: An Introduction* (Hoboken, NJ: John Wiley & Sons, 2020).

5. Doug McAdam, "Recruitment to High-Risk Activism: The Case of Freedom Summer," *American Journal of Sociology* 92, no. 1 (1986): 64–90, https://www.jstor.org/stable/2779717.

6. Hersh, *Politics Is for Power*.

7. Henrik Christensen, "Simply Slacktivism? Internet Participation in Finland," *E-journal of E-Democracy and Open Government* 4, no. 1 (2012): https://doi.org/10.29379/jedem.v4i1.93.

8. Pablo Barberá et al., "The Critical Periphery in the Growth of Social Protests," *PLOS One* 10, no. 11 (2015): e0143611, https://doi.org/10.1371/journal.pone.0143611; Alan Hamlin and Colin Jennings, "Expressive Political Behaviour: Foundations, Scope and Implications," *British Journal of Political Science* 41, no. 3 (2011): 645–70, https://doi:10.1017/S0007123411000020; Jose Marichal, "Political Facebook Groups: Micro-Activism and the Digital Front Stage," *First Monday* 18, no. 12–2 (2013), https://firstmonday.org/ojs/index.php/fm/article/download/4653/3800.

9. Ellie Brodie et al., *Pathways through Participation: What Creates and Sustains Active Citizenship?* (Brooklyn, NY: People Powered, 2011), https://drive.google.com/file/d/1G4d58iHAsKR3tcB0Dyl1BZtK5Sfe5LE4/view; Christensen, "Simply Slacktivism"; Hersh, *Politics Is for Power*; Carmen Leong et al., "Social Media Empowerment in Social Movements: Power Activation and Power Accrual in Digital Activism," *European Journal of Information Systems* 28, no. 2 (2019): 173–204, https://doi.org/10.1080/0960085X.2018.1512944; Marichal, "Political Facebook Groups."

10. Finger Lakes Climate Fund, accessed October 29, 2020, https://www.fingerlakesclimatefund.org/.

11. Gregg Sparkman and Gregory M. Walton, "Witnessing Change: Dynamic Norms Help Resolve Diverse Barriers to Personal Change," *Journal of Experimental Social Psychology* 82 (May 2019): 238–52, https://doi.org/10.1016/j.jesp.2019.01.007.

12. Paul C. Stern et al., "Values, Beliefs, and Proenvironmental Action: Attitude Formation toward Emergent Attitude Objects," *Journal of Applied Social Psychology* 25, no. 18 (1995): 1611–36, https://doi.org/10.1111/j.1559-1816.1995.tb02636.x.

13. Todd Rogers, Noah J. Goldstein, and Craig R. Fox, "Social Mobilization," *Annual Review of Psychology* 69, no. 1 (2018): 357–81, https://doi.org/10.1146/annurev-psych-122414-033718.

14. Andreas Nilsson, Magnus Bergquist, and P. Wesley Schultz, "Spillover Effects in Environmental Behaviors, across Time and Context: A Review and Research Agenda,"

Environmental Education Research 23, no. 4 (2017): 573–89, https://doi.org/10.1080/13504622.2016.1250148.

15. Damon Centola, *How Behavior Spreads: The Science of Complex Contagions* (Princeton, NJ: Princeton University Press, 2018).

16. Annie Lowrey, "All That Performative Environmentalism Adds Up," *The Atlantic*, August 31, 2020, https://www.theatlantic.com/ideas/archive/2020/08/your-tote-bag-can-make-difference/615817/.

17. Marianne E. Krasny et al., "E-Engagement: Approaches to Using Digital Communications in Student-Community Engagement," *Journal of Higher Education Outreach and Engagement* 25, no. 4 (2021): 21, https://openjournals.libs.uga.edu/jheoe/article/view/1727.

18. John D. McCarthy, "Persistence and Change among Nationally Federated Social Movements," in *Social Movements and Organizational Theory*, ed. Gerald F Davis et al. (Cambridge: Cambridge University Press, 2005), 193–225.

19. Brodie et al., *Pathways through Participation*; E. Gil Clary et al., "Understanding and Assessing the Motivations of Volunteers: A Functional Approach," *Journal of Personality and Social Psychology* 74, no. 6 (1998): 1516–30; Sarah-Louise Mitchell and Moira Clark, "Volunteer Choice of Nonprofit Organisation: An Integrated Framework," *European Journal of Marketing* 55, no. 1 (2020): 63–94, https://doi.org/10.1108/EJM-05-2019-0427.

20. Robin Zheng, "What Is My Role in Changing the System? A New Model of Responsibility for Structural Injustice," *Ethical Theory and Moral Practice* 21, no. 4 (2018): 869–85, https://doi.org/10.1007/s10677-018-9892-8.

21. Environmental Voter Project, accessed June 8, 2021, https://www.environmentalvoter.org/.

22. Elizabeth Cripps, "Individual Climate Justice Duties: The Cooperative Promotional Model and Its Challenges," in *Climate Justice and Non-State Actors: Corporations, Regions, Cities, and Individuals*, ed. Jeremy Moss and Lachlan Umbers (London: Routledge, 2020), 101–17.

23. Lance Bennett, Alexandra Segerberg, and Yunkang Yang, "The Strength of Peripheral Networks: Negotiating Attention and Meaning in Complex Media Ecologies," *Journal of Communication* 68, no. 4 (2018), https://doi.org/10.1093/joc/jqy032; Johan Enqvist, Maria Tengö, and Örjan Bodin, "Citizen Networks in the Garden City: Protecting Urban Ecosystems in Rapid Urbanization," *Landscape and Urban Planning* 130 (October 2014): 24–35, https://doi.org/10.1016/j.landurbplan.2014.06.007.

24. Kenneth T. Andrews and Michael Biggs, "The Dynamics of Protest Diffusion: Movement Organizations, Social Networks, and News Media in the 1960 Sit-Ins," *American Sociological Review* 71, no. 5 (2006): 752–77, http://www.jstor.org/stable/25472426.

25. Bennett et al., "Strength of Peripheral Networks"; Enqvist et al., "Citizen Networks in the Garden City"; Dana R. Fisher, Erika S. Svendsen, and James J. Connolly, *Urban Environmental Stewardship and Civic Engagement: How Planting Trees Strengthens the Roots of Democracy* (New York: Routledge, 2015), https://doi:10: 0415723639.

26. Barberá et al., "Critical Periphery"; Christensen, "Simply Slacktivism."

27. Enqvist et al., "Citizen Networks in the Garden City."

28. Laurel Besco, "Responses to the Clean Power Plan: Factors Influencing State Decision-Making," *Review of Policy Research* 35, no. 5 (2018): 670–90, https://doi.org/10.1111/ropr.12307; della Porta and Diani, *Social Movements*; Leong et al., "Social Media Empowerment."

29. Bennett et al., "Strength of Peripheral Networks"; Ronald S. Burt, "Reinforced Structural Holes," *Social Networks* 43 (October 2015): 149–61, https://doi.org/10.1016/j.

socnet.2015.04.008; Thomas W. Valente, "Network Interventions," *Science* 337, no. 6090 (2012): 49–53, https://doi.org/10.1126/science.1217330.

30. Andrews and Biggs, "Dynamics of Protest Diffusion."

31. Burt, "Reinforced Structural Holes"; Centola, *How Behavior Spreads*; Diani, "Social Movements and Collective Action"; Alexandra Segerberg, "Online and Social Media Campaigns for Climate Change Engagement," in *The Oxford Encyclopedia of Climate Change Communication*, ed. Matthew C. Nisbet et al. (Oxford: Oxford University Press, 2018), https://www.oxfordreference.com/view/10.1093/acref/9780190498986.001.0001/acref-9780190498986-e-398?rskey=WLO4CO&result=89.

32. Diani, "Social Movements and Collective Action."

33. Megan E. Brooker and David S. Meyer, "Coalitions and the Organization of Collective Action," in *The Wiley Blackwell Companion to Social Movements*, ed. David A Snow et al. (Hoboken, NJ: Wiley Blackwell, 2018), 252–68; Enqvist et al., "Citizen Networks in the Garden City"; Henrik Ernstson, Sverker Sorline, and Thomas Elmqvist, "Social Movements and Ecosystem Services: The Role of Social Network Structure in Protecting and Managing Urban Green Areas in Stockholm," *Ecology and Society* 13, no. 2 (2008): 39, http://www.ecologyandsociety.org/vol13/iss2/art39/; Jiawei Sophia Fu and Katherine R. Cooper, "Interorganizational Network Portfolios of Nonprofit Organizations: Implications for Collaboration Management," *Nonprofit Management and Leadership* 31 (2020): 437–59, https://doi.org/10.1002/nml.21438; James Kitts, "Mobilizing in Black Boxes: Social Networks and Participation in Social Movement Organizations," *Mobilization: An International Quarterly* 5, no. 2 (2000): 241–57, https://doi.org/10.17813/maiq.5.2.5408016w34215787; Sarah A. Soule and Conny Roggeband, "Diffusion Processes within and across Movements," in Snow, Soule, and Kriesi, *Wiley Blackwell Companion to Social Movements*, 236–51.

34. Mario Diani, "Networks and Social Movements," in *The Blackwell Encyclopedia of Sociology*, ed. George Ritzer (Hoboken, NJ: Wiley-Blackwell, 2015), https://doi.org/10.1002/9781405165518.wbeoss162.pub2; Erich Steinman, "Why Was Standing Rock and the #Nodapl Campaign So Historic? Factors Affecting American Indian Participation in Social Movement Collaborations and Coalitions," *Ethnic and Racial Studies* 42, no. 7 (2019): 1070–90, https://doi.org/10.1080/01419870.2018.1471215.

35. Rebecca Solnit, "Standing Rock Inspired Ocasio-Cortez to Run. That's the Power of Protest," *The Guardian*, January 14, 2019, https://www.theguardian.com/commentisfree/2019/jan/14/standing-rock-ocasio-cortez-protest-climate-activism.

36. Philip Oltermann, "The Wurst Is Over: Why Germany Now Loves to Go Vegetarian," *The Guardian*, September 27, 2020, https://www.theguardian.com/world/2020/sep/27/the-wurst-is-over-why-germany-land-of-schnitzels-now-loves-to-go-vegetarian.

37. Justin McCarthy and Scott Dekoster, "Nearly One in Four in U.S. Have Cut Back on Eating Meat," Gallup, January 27, 2020, https://news.gallup.com/poll/282779/nearly-one-four-cut-back-eating-meat.aspx.

38. Daniel Welch and Luke Yates, "The Practices of Collective Action: Practice Theory, Sustainability Transitions and Social Change," *Journal for the Theory of Social Behaviour* 48, no. 3 (2018): 288–305, https://doi.org/10.1111/jtsb.12168.

39. Albert Bandura, "Self-Efficacy: Toward a Unifying Theory of Behavioral Change," *Psychological Review* 84, no. 2 (1977): 191–215, https://doi.org/10.1037/0033-295X.84.2.191.

40. Martijn van Zomeren, Tamar Saguy, and Fabian M. H. Schellhaas, "Believing in 'Making a Difference' to Collective Efforts: Participative Efficacy Beliefs as a Unique Predictor of Collective Action," *Group Processes and Intergroup Relations* 16, no. 5 (2013): 618–34, https://doi.org/10.1177/1368430212467476.

41. "Albert Bandura's Influence on the Field of Psychology," Verywell Mind, March 15, 2020, https://www.verywellmind.com/albert-bandura-biography-1925-2795537.

42. Bandura, "Self-Efficacy: Toward a Unifying Theory of Behavioral Change"; Albert Bandura, *Self-Efficacy: The Exercise of Control* (New York: W. H. Freeman, 1997).

43. Elizabeth Beaumont, "Political Agency and Empowerment: Pathways for Developing a Sense of Political Efficacy in Young Adults," in *Handbook of Research on Civic Engagement in Youth*, ed. Lonnie Sherrod, Judith Torney-Purta, and Constance Flanagan (Somerset, NJ: John Wiley & Sons, 2010), 525–58.

44. Bandura, "Self-Efficacy."

45. Bandura, *Self-Efficacy*.

46. Van Zomeren et al., "Believing in 'Making a Difference.'"

47. Beaumont, "Political Agency and Empowerment."

48. Mattha Busby, "Extinction Rebellion Activists Stage Die-in Protests across Globe," *The Guardian*, April 27, 2019, https://www.theguardian.com/environment/2019/apr/27/extinction-rebellion-activists-stage-die-in-protests-across-globe.

49. Beaumont, "Political Agency and Empowerment," 554.

50. Bandura, "Self-Efficacy"; Beaumont, "Political Agency and Empowerment."

51. "Ayana Elizabeth Johnson and Naomi Oreskes: The Schneider Award," December 30, 2021, in *Climate One*, podcast, https://www.climateone.org/audio/ayana-elizabeth-johnson-and-naomi-oreskes-schneider-award.

52. "Albert Bandura's Influence."

53. Simon Hattenstone, "The Transformation of Greta Thunberg," *The Guardian*, September 25, 2021, https://www.theguardian.com/environment/ng-interactive/2021/sep/25/greta-thunberg-i-really-see-the-value-of-friendship-apart-from-the-climate-almost-nothing-else-matters.

54. Mike De Socio, "The U.S. City That Has Raised $100M to Climate-Proof Its Buildings," *The Guardian*, August 19, 2021, https://www.theguardian.com/environment/2021/aug/19/ithaca-new-york-raised-100m-climate-proof-buildings.

55. Erica Chenoweth, *Civil Resistance: What Everyone Needs to Know* (New York: Oxford University Press, 2021).

56. Chloe Cockburn, "Philanthropists Must Invest in an Ecology of Change," *Stanford Social Innovation Review*, June 25, 2018, https://doi.org/10.48558/VWG3-3331; alexrjl, "Why I'm Concerned about Giving Green," Effective Altruism Forum, accessed January 1, 2022, https://forum.effectivealtruism.org/posts/7yN7SKPpL3zN7yfcM/why-i-m-concerned-about-giving-green.

57. Della Porta and Diani, *Social Movements*.

58. Clive Barnett et al., *Globalizing Responsibility: The Political Rationalities of Ethical Consumption* (West Sussex: Wiley-Blackwell, 2011); Michele Micheletti and Dietlind Stolle, "Consumer Strategies in Social Movements," in *The Oxford Handbook of Social Movements*, ed. Donatella della Porta and Mario Diani (Oxford: Oxford University Press, 2015), 10.1093/oxfordhb/9780199678402.013.45.

59. Della Porta and Diani, *Social Movements*, 49.

60. Rebecca Willis, "Constructing a 'Representative Claim' for Action on Climate Change: Evidence from Interviews with Politicians," *Political Studies* 66, no. 4 (2018): 940–58, https://doi.org/10.1177/0032321717753723.

61. Robert J Sampson et al., "Civil Society Reconsidered: The Durable Nature and Community Structure of Collective Civic Action," American Journal of Sociology 111, no. 3 (2005): 673–714, https://doi.org/10.1086/497351.

62. Della Porta and Diani, *Social Movements*, 49.

63. Chenoweth, *Civil Resistance*.

64. James J. Patterson, *Remaking Political Institutions: Climate Change and Beyond* (Cambridge, UK: Cambridge University Press, 2020), doi:10.1017/9781108769341.

65. Somini Sengupta, "Big Setbacks Propel Oil Giants toward a 'Tipping Point,'" *New York Times*, May 29, 2021, https://www.nytimes.com/2021/05/29/climate/fossil-fuel-

courts-exxon-shell-chevron.html?action=click&module=Well&pgtype=Homepage§ion=Climate%20and%20Environment.

66. Hiroko Tabuchi, "In Your Facebook Feed: Oil Industry Pushback against Biden Climate Plans," *New York Times*, September 30, 2021, https://www.nytimes.com/2021/09/30/climate/api-exxon-biden-climate-bill.html.

67. Oliver Milman, "Apple and Disney among Companies Backing Groups against U.S. Climate Bill," *The Guardian*, October 1, 2021, https://www.theguardian.com/us-news/2021/oct/01/apple-amazon-microsoft-disney-lobby-groups-climate-bill-analysis.

68. Karla Adam and Harry Stevens, "Who Has the Most Delegates at the COP26 Summit? The Fossil Fuel Industry," *Washington Post*, November 8, 2021, https://www.washingtonpost.com/world/2021/11/08/cop26-glasgow-climate-summit-fossil-fuel/.

69. Aaron Gregg, "Shell to Move Headquarters to U.K., Revamp Share Structure and Drop 'Royal Dutch,'" *Washington Post*, November 15, 2021, https://www.washingtonpost.com/business/2021/11/15/royal-dutch-shell-netherlands-uk/.

70. Johan Rockström et al., "Planetary Boundaries: Exploring the Safe Operating Space for Humanity," *Ecology and Society* 14, no. 2 (2009): 32, http://www.ecologyandsociety.org/vol14/iss2/art32/.

71. Ilona M. Otto et al., "Social Tipping Dynamics for Stabilizing Earth's Climate by 2050," *Proceedings of the National Academy of Sciences* 117, no, 5 (2020): 2354–65, https://doi.org/10.1073/pnas.1900577117.

72. Ana M. Aranda and Tal Simons, "On Two Sides of the Smoke Screen: How Activist Organizations and Corporations Use Protests, Campaign Contributions, and Lobbyists to Influence Institutional Change," in *Social Movements, Stakeholders and Non-Market Strategy*, ed. Forest Briscoe, Brayden G. King, and Jocelyn Leitzinger (Bingley, UK: Emerald Publishing Limited, 2018), 261–315.

73. Leong et al., "Social Media Empowerment."

74. De Socio, "U.S. City That Has Raised $100M."

75. Christopher Marquis, *Better Business: How the B Corp Movement Is Remaking Capitalism* (Ithaca, NY: Christoper Marquis, 2020).

76. "A Vision for Equitable Climate Action," U.S. Climate Action Network, accessed November 23, 2020, https://equitableclimateaction.org/.

77. "THRIVE Agenda," Green New Deal Network, accessed November 23, 2020, https://www.thriveagenda.com/.

78. "Action for Climate Empowerment," United Nations, accessed November 23, 2020, https://unfccc.int/topics/education-youth/the-big-picture/what-is-action-for-climate-empowerment.

79. "Biden Plan for a Clean Energy Revolution and Environmental Justice," Biden for President, accessed November 23, 2020, https://joebiden.com/climate-plan/#.

80. Robin Wall Kimmerer, *Braiding Sweetgrass: Indigenous Wisdom, Scientific Knowledge and the Teachings of Plants* (Minneapolis: Milkweed Editions, 2013).

81. Paul Engler, "Protest Movements Need the Funding They Deserve," *Stanford Social Innovation Review*, July 3, 2018, https://doi.org/10.48558/AYM1-SK19.

82. Chenoweth, *Civil Resistance.*

5. RESPONSIBILITY

1. Paul Griffin, *The Carbon Majors Database: CDP Carbon Majors Report 2017*, July 2017, https://cdn.cdp.net/cdp-production/cms/reports/documents/000/002/327/original/Carbon-Majors-Report-2017.pdf?1501833772; Richard Heede, *Carbon Majors: Update of Top Twenty Companies 1965–2017*, Climate Accountability Institute, October 9, 2019, https://climateaccountability.org/pdf/CAI%20PressRelease%20Top20%20Oct19.pdf; Naomi Oreskes, "Playing Dumb on Climate Change," *New York Times*, January 3,

2015, http://www.nytimes.com/2015/01/04/opinion/sunday/playing-dumb-on-climate-change.html?hp&action=click&pgtype=Homepage&module=c-column-top-span-region®ion=c-column-top-span-region&WT.nav=c-column-top-span-region&_r=0; Naomi Oreskes and Erik M. Conway, *Merchants of Doubt: How a Handful of Scientists Obscured the Truth on Issues from Tobacco Smoke to Climate Change* (New York: Bloomsbury, 2011); Geoffrey Supran and Naomi Oreskes, "Rhetoric and Frame Analysis of ExxonMobil's Climate Change Communications," *One Earth* 4, no. 5 (2021): 696–719, https://doi.org/10.1016/j.oneear.2021.04.014; Amy Westervelt, "Campaigns So Successful They've Landed in Court," in *Drilled: Critical Frequency*, podcast, November 17, 2017, https://podcasts.google.com/feed/aHR0cHM6Ly9mZWVkcy5tZWdhcGhvbmUuZm0vQ0ZRWTIyOTE3NzIzOTU/episode/MDlhNDcyZjNlODQyNDM1NjkxMTQ1Mjg5MjE5ZThlOGQ?sa=X&ved=0CAUQkfYCahgKEwig5IObt571AhUAAAAAHQAAAAAQ-RA&hl=en.

2. Benjamin Franta, "Weaponizing Economics: Big Oil, Economic Consultants, and Climate Policy Delay," *Environmental Politics*, August 21, 2021, 1–21, https://doi.org/10.1080/09644016.2021.1947636.

3. Franta, "Weaponizing Economics."

4. 151 Cong. Rec. S7015 (2005), 37, https://www.govinfo.gov/content/pkg/CREC-2005-06-22/pdf/CREC-2005-06-22-pt1-PgS6980-4.pdf#page=68.

5. Kat Kramer and Joe Ware, *Counting the Cost 2021: A Year of Climate Breakdown* (London: Christian Aid, 2021), https://app.box.com/s/ui6b821a8x38uby54i7hfsqm-3fadyzx8/file/896537370689.

6. Franta, "Weaponizing Economics."

7. Amy Westervelt, "Exploiting Scientists' Kryptonite: Certainty," in *Drilled: Critical Frequency*, podcast, November 17, 2016, https://www.spreaker.com/user/15244480/exploiting-scientists-kryptonite-certain.

8. "The Project," Count Us In, accessed October 15, 2020, https://www.count-us-in.org/.

9. Fiona Harvey, "Campaign Seeks 1Bn People to Save Climate—One Small Step at a Time," *The Guardian*, October 10, 2020, https://www.theguardian.com/environment/2020/oct/10/campaign-seeks-1bn-people-to-save-climate-one-small-step-time.

10. Dino Grandoni, "States and Cities Scramble to Sue Oil Companies over Climate Change," *Washington Post*, September 14, 2020, https://www.washingtonpost.com/climate-environment/2020/09/14/states-cities-scramble-sue-oil-companies-over-climate-change/.

11. Dario Kenner, *Carbon Inequality: The Role of the Richest in Climate Change* (New York: Routledge, 2019); Laui Lahikainen. "Individual Responsibility for Climate Change: A Social Structural Account" (PhD diss., University of Tampere, 2018).

12. Robert H. Frank, *Under the Influence: Putting Peer Pressure to Work* (Princeton, NJ: Princeton University Press, 2020); "Ipsos Encyclopedia—Affluent," Ipsos, accessed November 23, 2020, https://www.ipsos.com/en/ipsos-encyclopedia-affluent.

13. Elizabeth Cripps, "Individual Climate Justice Duties: The Cooperative Promotional Model and Its Challenges," in *Climate Justice and Non-State Actors: Corporations, Regions, Cities, and Individuals*, ed. Jeremy Moss and Lachlan Umbers (London: Routledge, 2020), 101–17; Marion Smiley, "Collective Responsibility," in *Stanford Encyclopedia of Philosophy* (Stanford, CA: Stanford University, 1997–), article last modified March 27, 2017, https://plato.stanford.edu/archives/sum2017/entries/collective-responsibility/; Matthew Talbert, "Moral Responsibility," in *Stanford Encyclopedia of Philosophy*, article published October 16, 2019, https://plato.stanford.edu/archives/win2019/entries/moral-responsibility/.

14. Josh Levin, *The Queen: The Forgotten Life behind an American Myth* (New York: Little, Brown, 2019).

15. Iris Marion Young, *Responsibility for Justice* (Oxford: Oxford University Press, 2011).

16. "What Are Covenants?," Mapping Prejudice, Borchert Map Library, University of Minnesota, accessed January 21, 2022, https://mappingprejudice.umn.edu/what-are-covenants/.

17. Neal Tognazzini and D. Justin Coates, "Blame," in *Stanford Encyclopedia of Philosophy*, article last modified August 17, 2018, https://plato.stanford.edu/archives/sum2021/entries/blame/.

18. Tognazzini and Coates, "Blame."

19. Tognazzini and Coates, "Blame."

20. Joy Smithson and Steven Venette, "Stonewalling as an Image-Defense Strategy: A Critical Examination of BP's Response to the Deepwater Horizon Explosion," *Communication Studies* 64, no. 4 (2013): 395–410, https://doi.org/10.1080/10510974.2013.770409.

21. Sabine Matejek and Tobias Gössling, "Beyond Legitimacy: A Case Study in BP's 'Green Lashing,'" *Journal of Business Ethics* 120, no. 4 (2014): 571–84, https://doi.org/10.1007/s10551-013-2006-6.

22. Steven Mufson, "After the Deepwater Horizon Disaster, a New BP Emerges," *Anchorage Daily News*, July 16, 2018, https://www.adn.com/business-economy/energy/2018/07/16/after-deepwater-horizon-a-new-bp-emerges/.

23. David Uhlmann, "BP Paid a Steep Price for the Gulf Oil Spill but for the US a Decade Later, It's Business as Usual," *The Conversation*, April 23, 2020, https://theconversation.com/bp-paid-a-steep-price-for-the-gulf-oil-spill-but-for-the-us-a-decade-later-its-business-as-usual-136905.

24. Rebecca Leber, "What the Oil Industry Still Won't Tell Us," Vox, October 28, 2021, https://www.vox.com/22745597/big-oil-congress-hearing-exxonmobil-bp-chevron-shell.

25. Maxine Joselow and Dino Grandoni, "Big Oil CEOs Testify before House Oversight Committee," *Washington Post*, October 28, 2021, https://www.washingtonpost.com/climate-environment/2021/10/28/oil-executives-testimony-live-updates/.

26. Leber, "What the Oil Industry Still Won't Tell Us."

27. Hiroko Tabuchi and Lisa Friedman, "Oil Executives Grilled over Industry's Role in Climate Disinformation," *New York Times*, October 28, 2021, https://www.nytimes.com/2021/10/28/climate/oil-executives-house-disinformation-testimony.html.

28. Tabuchi and Friedman, "Oil Executives Grilled."

29. Joselow and Grandoni, "Big Oil CEOs Testify."

30. Robin Zheng, "What Is My Role in Changing the System? A New Model of Responsibility for Structural Injustice," *Ethical Theory and Moral Practice* 21, no. 4 (2018): 882, https://doi.org/10.1007/s10677-018-9892-8.

31. Behnam Taebi and Azar Safari, "On Effectiveness and Legitimacy of 'Shaming' as a Strategy for Combatting Climate Change," *Science and Engineering Ethics* 23, no. 5 (2017): 1289–306, https://doi.org/10.1007/s11948-017-9909-z.

32. John Cook et al. *America Misled: How the Fossil Fuel Industry Deliberately Misled Americans about Climate Change* (Fairfax, VA: George Mason University Center for Climate Change Communication, 2019), https://www.climatechangecommunication.org/america-misled/; Oreskes and Conway, *Merchants of Doubt*; Supran and Oreskes, "Rhetoric and Frame Analysis of Exxonmobil's Climate Change Communications."

33. Jennifer Jacquet, *Is Shame Necessary?* (New York: Pantheon Books, 2015).

34. Amrit Chaudhari, *Greenpeace, Nestlé and the Palm Oil Controversy: Social Media Driving Change?* (Hyderabad, India: Center for Management Research, 2011).

35. Smiley, "Collective Responsibility"; Talbert, "Moral Responsibility."

36. Elizabeth Cripps, *Climate Change and the Moral Agent: Individual Duties in an Interdependent World* (Oxford: Oxford University Press, 2013).

37. Cripps, *Climate Change and the Moral Agent*; Smiley, "Collective Responsibility."

38. Simon Caney, "Climate Justice," in *Stanford Encyclopedia of Philosophy*, article published June 4, 2020, https://plato.stanford.edu/archives/win2021/entries/justice-climate/.

39. Cripps, "Individual Climate Justice Duties."

40. Cripps, *Climate Change and the Moral Agent.*

41. Young, *Responsibility for Justice.*

42. Lahikainen, "Individual Responsibility for Climate Change."

43. Maeve McKeown, "Iris Marion Young's 'Social Connection Model' of Responsibility: Clarifying the Meaning of Connection," *Journal of Social Philosophy* 49, no. 3 (2018): 485, https://doi.org/10.1111/josp.12253.

44. McKeown, "'Social Connection Model,'"500, italics in original.

45. Iris Marion Young, "Responsibility and Global Justice: A Social Connection Model," *Social Philosophy and Policy* 23, no. 1 (2006): 102–30, https://doi.org/10.1017/S0265052506060043; Young, *Responsibility for Justice.*

46. Young, *Responsibility for Justice*, 93.

47. Zheng, "What Is My Role?"

48. Nikki Fortier, "ETMP Discussion of Robin Zheng's 'What Is My Role in Changing the System?': A New Model of Responsibility for Structural Injustice," Pea Soup, June 29, 2018, http://peasoup.us/2018/06/etmp-discussion-of-robin-zhengs-what-is-my-role-in-changing-the-system-a-new-model-of-responsibility-for-structural-injustice/; Zheng, "What Is My Role?"

49. Susan Clayton, "Environment and Identity," in *The Oxford Handbook of Environmental and Conservation Psychology*, ed. Susan Clayton (Oxford: Oxford University Press, 2012), 164–80; Patrick Devine-Wright and Susan Clayton, "Introduction to the Special Issue: Place, Identity and Environmental Behaviour," *Journal of Environmental Psychology* 30, no. 3 (2010): 267–70, https://doi.org/10.1016/S0272-4944(10)00078-2.

50. Zheng, "What Is My Role?"

51. Anthony Giddens, "Elements of the Theory of Structuration," in *Practicing History: New Directions in Historical Writing after the Linguistic Turn*, ed. Gabrielle Spiegel (New York: Routledge, 2005), 121–42.

52. Nick Nash et al., "Climate-Relevant Behavioral Spillover and the Potential Contribution of Social Practice Theory," *Wiley Interdisciplinary Reviews: Climate Change* 8, no. 6 (2017): e481, https://doi.org/10.1002/wcc.481; Nick Nash et al., "Reflecting on Behavioral Spillover in Context: How Do Behavioral Motivations and Awareness Catalyze Other Environmentally Responsible Actions in Brazil, China, and Denmark?," *Frontiers in Psychology* 10, no. 788 (2019), https://doi.org/10.3389/fpsyg.2019.00788; Gregg Sparkman, Shahzeen Attari, and Elke Weber, "Moderating Spillover: Focusing on Personal Sustainable Behavior Rarely Hinders and Can Boost Climate Policy Support," *Energy Research and Social Science* 78 (August 2021), https://doi.org/10.1016/j.erss.2021.102150.

53. Zheng, quoted in Fortier, "ETMP Discussion."

54. Zheng, "What Is My Role?"

55. Robert Cox, "Nature's 'Crisis Disciplines': Does Environmental Communication Have an Ethical Duty?," *Environmental Communication* 1, no. 1 (2007): 5–20, https://doi.org/10.1080/17524030701333948.

56. Fortier, "ETMP Discussion."

57. Zheng, "What Is My Role?," 880, italics in original.

58. Daniel Hunter, *Building a Movement to End the New Jim Crow: An Organizing Guide* (Denver: Veterans of Hope, 2015).

59. Daniel Kreiss and Zeynep Tufekci, "Occupying the Political: Occupy Wall Street, Collective Action, and the Rediscovery of Pragmatic Politics," *Cultural Studies ↔ Critical Methodologies* 13, no. 3 (2013): 163–67, https://doi.org/10.1177/1532708613477367.

60. Alice Malpass et al., "Problematizing Choice: Responsible Consumers and Sceptical Citizens," in *Governance, Consumers and Citizens: Consumption and Public Life*, ed. Mark Bevir and Frank Trentmann (London: Palgrave Macmillan, 2007), 247.

61. Zheng, "What Is My Role?"

62. "The Guardian Launches New Green Initiatives Including Climate Data Dashboard and Outlines Plans for How It Will Achieve Net Zero Emissions by 2030," *The Guardian*, October 5, 2020, https://www.theguardian.com/gnm-press-office/2020/oct/05/the-guardian-launches-new-green-initiatives-including-climate-data-dashboard-and-outlines-plans-for-how-it-will-achieve-net-zero-emissions-by-2030.

63. Allan Jenkins, "The 20 Best Sandwich Recipes," *The Guardian*, October 26, 2020, https://www.theguardian.com/food/2020/oct/26/the-20-best-sandwich-recipes.

AFTERWORD

1. Kristian Nielsen et al., "How Psychology Can Help Limit Climate Change," *American Psychologist* 76, no. 1 (2020): 130–44, https://doi.org/10.1037/amp0000624.

Index